LAST
STAND

Smithsonian Books

Collins

LAST
STAND

George Bird Grinnell,
the Battle to Save the Buffalo,
and the Birth of the New West

MICHAEL PUNKE

HarperCollins books may be purchased for educational, business, or sales promotional
use. For information please write: Special Markets Department, HarperCollins
Publishers Inc., 10 East 53rd Street, New York, NY 10022.
First Smithsonian Books edition published 2007.

Designed by Janet M. Evans

The Library of Congress Cataloging-in-Publication Data has been
applied for.

ISBN 978-0-06-089782-6

07 08 09 10   ID/QW   10 9 8 7 6 5 4 3 2 1

LAST
STAND

George Bird Grinnell,
the Battle to Save the Buffalo,
and the Birth of the New West

MICHAEL PUNKE

HarperCollins books may be purchased for educational, business, or sales promotional
use. For information please write: Special Markets Department, HarperCollins
Publishers Inc., 10 East 53rd Street, New York, NY 10022.
First Smithsonian Books edition published 2007.

Designed by Janet M. Evans

The Library of Congress Cataloging-in-Publication Data has been
applied for.

ISBN 978-0-06-089782-6

07 08 09 10   ID/QW   10 9 8 7 6 5 4 3 2 1

For Sophie and Bo:

May your children's children see wild buffalo on the plains.

Motionless, with head thrown back,
and in an attitude of attention,
he calmly inspected the vessel floating along below him;
so beautiful an object amid his wild surroundings,
and with his background of brilliant sky,
that no hand was stretched out for the rifle ...

There is one spot left,
a single rock about which this tide will break,
and past which it will sweep, leaving it undefiled
by the unsightly traces of civilization.

George Bird Grinnell

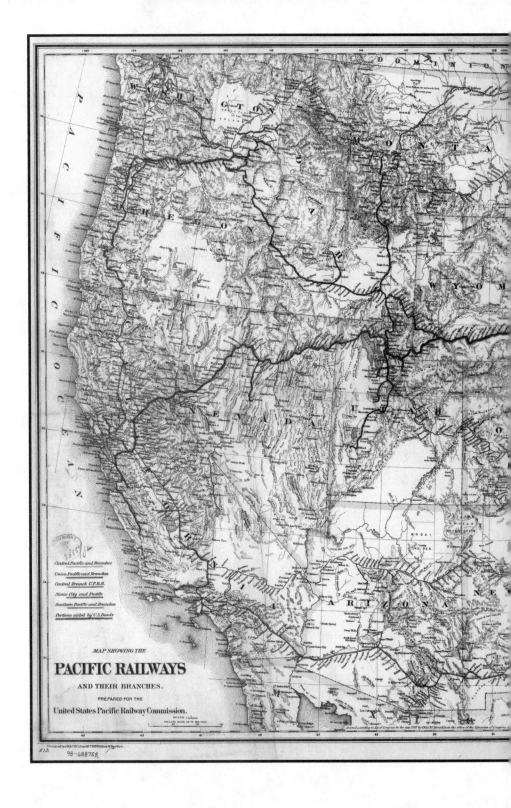

MAP SHOWING THE

PACIFIC RAILWAYS

AND THEIR BRANCHES.

PREPARED FOR THE

United States Pacific Railway Commission.

Central Pacific and Branches
Union Pacific and Branches
Central Branch U.P.R.R.
Sioux City and Pacific
Southern Pacific and Branches
Portions aided by U.S.Roads

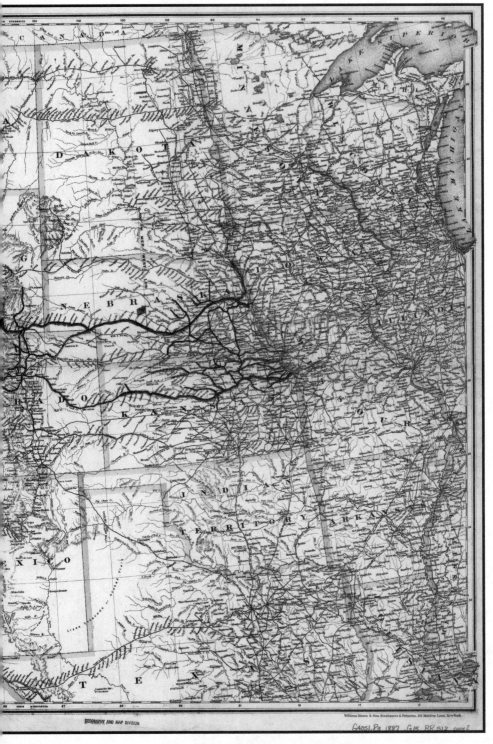

The West of the Buffalo and George Bird Grinnell.
Courtesy of the Library of Congress, Geography and Map Division.

CONTENTS

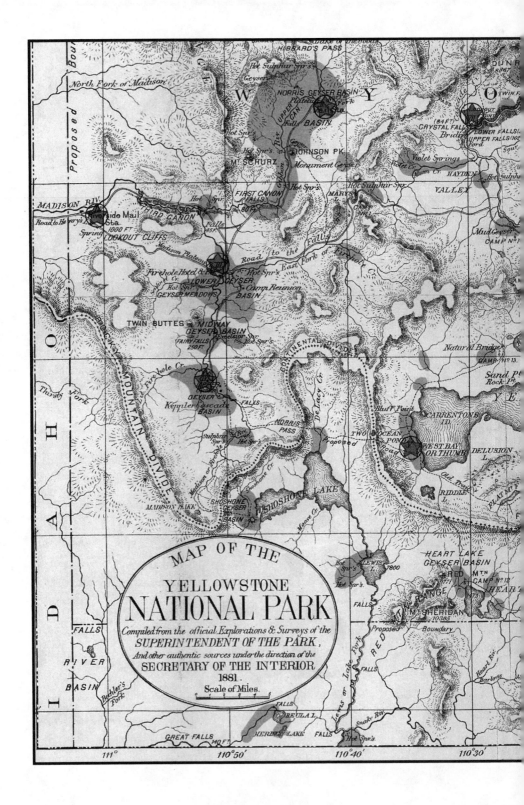

MAP OF THE
YELLOWSTONE
NATIONAL PARK
Compiled from the official Explorations & Surveys of the
SUPERINTENDENT OF THE PARK,
And other authentic sources under the direction of the
SECRETARY OF THE INTERIOR
1881.
Scale of Miles.

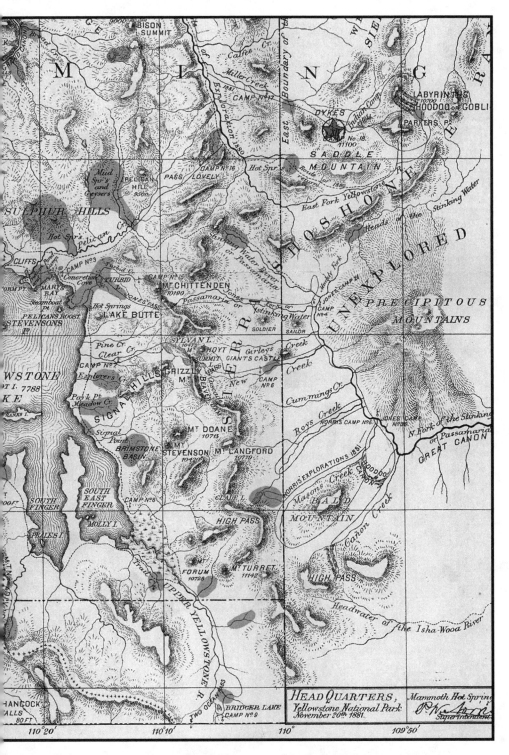

Yellowstone National Park, 1881

Courtesy of the Library of Congress, Geography and Map Division.

THE STAND

*After placing about fifteen shots where they were most
needed, I had the herd stopped, and the buffalo paid no
attention to the subsequent shooting.*

—Victor Grant Smith

Vic Smith, a hunter, lifted his head above a rise on the plains floor, peering down at several hundred buffalo in the valley of the Redwater River. The Montana winter of 1881 was frigid, all the more so because Smith lay prone in the snow, two Sharps buffalo rifles and several bandoleers of cartridges spread out on a tarp beside him. Smith was careful to stay downwind and wore a white sheet to conceal him from the nearest animals, three hundred yards away. For a while he just watched, his experienced eyes studying the herd—picking out the leaders, anticipating movements, carefully planning his first shot. It all looked perfect, the ideal stand.

Finally Smith reached for one of the Sharps, working the lever to chamber a four-inch brass shell. Supporting the stout barrel across his arm, Smith sighted carefully on the old cow that he knew led the herd. He aimed at a spot just in front of her hip, then fired.

The report of the big gun thundered across the wide plain, and a cloud of acrid smoke temporarily obscured the herd. Smith did not look to see if he had hit his target—he knew he had. Instead he set the smoking rifle on the tarp and loaded the second gun, then pulled it snug to his shoulder. He alternated rifles each shot; otherwise the barrels became so hot that they fouled. In Texas, he'd heard, buffalo hunters sometimes urinated on their guns to cool them, but in Montana, winter did the work.

The second Sharps ready, Smith looked up to find exactly what he expected. His first shot had found its precise target in front of the cow's hip. When hit in that spot, Smith knew, the animal could not run off but instead would just stand there, "all humped up with pain." As Smith intended, other members of the herd—the old cow's "children, grand-children, cousins, and aunts"—were already starting to mill about, confused, some sniffing at the blood that seeped from the cow.

Smith now sighted on another old cow on the opposite side of the herd, marking the same target in front of the hip. He fired again.

Smith worked deliberately, never rushing, a shot about once every thirty seconds. Every bullet was strategic. Most of the early targets were cows, though occasionally he picked off a skittish bull that looked ready to bolt. "After placing about fifteen shots where they were most needed," he would later recall, "I had the herd stopped, and the buffalo paid no attention to the subsequent shooting." Experienced hunters like Smith called it "tranquilizing" or "mesmerizing" the herd.

An hour later he was done. Below Vic Smith in the valley of the Redwater lay 107 dead buffalo. In the 1881 season he would kill 4,500.[1]

THE STORY OF HOW THE BUFFALO WAS SAVED FROM EXTINCTION IS one of the great dramas of the Old West. More profoundly, it is a story of the transition from the Old West to the New—a transition whose battles are still fought bitterly to this day. The story is personified in a man, little known today, by the name of George Bird Grinnell. Grinnell was a scientist and a journalist, a hunter and a conservationist. In his remarkable life, Grinnell would live the adventures of the Old West even as he helped to shape the New.

CHAPTER ONE

"Wild and Wooly"

*The party started from New Haven late in June, bound
for a West that was then really wild and wooly.*

—GEORGE BIRD GRINNELL,
Memories

T he adventure that changed the course of George Bird Grinnell's
life began with a train, and the path of the train, as it crossed the
plains in the summer of 1870, was blocked by buffalo.

The new transcontinental railroad, like the wagon trails that preceded
it, hewed to the valleys. Far from "featureless," as the Great Plains is frequently described, it is a region whose signature characteristic is so pervasive as to overwhelm—an openness so vast that the newcomer has no
antecedent to place it in context. Coming, as Grinnell did, from the East,
with its hemmed-in horizons and creeping green, arrival on the stark
prairie was a shock to the system, an obvious demarcation of a place that
was new. It was also, in the summer of 1870, a place that was wild.

As the train glided along the tracks, Grinnell heard the sudden screech of
metal brakes and excited shouts. Looking out the window, he saw a herd of
buffalo. After a brief delay, the herd wandered off and the voyage continued.
Later, though, the train was halted a second time by another herd. "We supposed they would soon pass by," remembered Grinnell, "but they kept coming
. . . in numbers so great that they could not be computed." It took *three hours*
for the herd to cross the tracks.[1] In the early days of the railroad, the problem
of buffalo blocking tracks was so common that engines were sometimes
equipped with a device that shot out steam to scatter the herd.

For the nineteenth-century traveler, no sight better symbolized arrival in the West than the buffalo. Grinnell, who would turn twenty-one in two months, had arrived in the midst of his boyhood dreams. He certainly spoke volumes about his own motivations when he later wrote that "none of [us] except the leader had any motive for going other than the hope of adventure with wild game or wild Indians."[2]

Grinnell and his young companions certainly looked prepared for adventure. Each of the young men carried a shiny new Henry repeating rifle, a pistol, bandoleers of cartridges, and a Bowie knife. Never mind that few had any experience with weapons (Grinnell was one who did). In Omaha, they had walked out onto the prairie "to try our fire arms." Grinnell, at least, was under no illusion: "The members of the party were innocent of any knowledge of the western country, but its members pinned their faith to Professor Marsh."[3]

"We supposed they would soon pass by, but they kept coming . . .
in numbers so great that they could not be computed."
A Hold-Up on the Kansas-Pacific, 1869, by Martin Garretson.
Courtesy of the National Museum of Wildlife Art.

Eyes to the West: George Bird
Grinnell in his early twenties.
Courtesy of the Scott Meyer family.

AMERICA OF THE NINETEENTH CENTURY LACKED ROYALTY, BUT IT
was not without aristocracy, and the family of George Bird Grinnell had
bequeathed to him a station near the uppermost strata. Young George
could trace his pedigree to the *Mayflower*. Indeed his ancestors included
Betty Alden, immortalized by Jane G. Austin in her book *Betty Alden:
The First-Born Daughter of the Pilgrims*. Grinnell's forefathers had been
leading Americans since long before the United States came into being.
Five had served as colonial governors. His grandfather, George Grinnell,
served ten terms as a U.S. congressman.[4]

George Bird Grinnell was born on September 20, 1849, in Brooklyn,
the first of five children to Helen A. Lansing and George Blake Grinnell.
Grinnell's father began his career as a successful dry-goods merchant
and ended it as a prominent merchant banker—the "principal agent in
Wall Street of Commodore Cornelius Vanderbilt."[5]

As a young George Bird Grinnell contemplated his future, the path of least
resistance seemed to flow naturally toward a position as a captain of finance in
a world ruled by the class to which he was born. Certainly this was the direction that his father and mother would push. Instead Grinnell would one day
rise to challenge the foundational tenets on which his world had been built.

The events that put Grinnell on a different course began on New Year's Day, 1857. He was 7 that year, and his father moved the family to the country. They rented at first, eventually building a house on a large tract of land in a part of Manhattan known as Audubon Park. The entire area once had been owned by John James Audubon, the famous painter-naturalist. Today, the quarter has been swallowed whole by New York City, bounded by West 158th and West 155th streets to the north and south, the Hudson River and Amsterdam Avenue to the east and west. In 1857, though, New York City was far away. Access to the city was by the Hudson River Railroad or by wagon, a trip of one and a half hours over hilly terrain.

Though John James Audubon had been dead for six years when the Grinnells moved to Audubon Park, much of the artist's family was still in residence. Audubon's two adult sons, Gifford and Woodhouse, continued the painting and publishing enterprise of their father. Each had a family and a house of his own on the property. Lucy Audubon, the elderly widow of the artist, lived with Gifford.

For a young boy, Audubon Park was an idyllic playground, like living in an engraving from Currier & Ives. "In the early days of Audubon Park almost nothing was seen of what in later days was called 'improvement,'" as Grinnell later described it. "The fields and woods were left in a state of nature." There were great groves of hemlock, chestnut, and oak. Springs flowed up from the ground and brooks tumbled down to the Hudson. There were stables with horses, pens of cattle and pigs, free-roaming chickens, geese, and ducks. The land was wild enough to be thick with small game, songbirds, and birds of prey, and Grinnell remembered a time when three eagles fought for a fish on his front lawn.[6]

Between the Grinnells, the Audubons, and the handful of other families who inhabited Audubon Park, a veritable tribe of young children roamed the whole area at will. Grinnell's boyhood memories, which he recorded in an unpublished document for his nieces and nephews, read like *Tom Sawyer*—an almost mythic childhood of unsupervised adventure and delightfully harmless delinquency. Grinnell and his cohorts stalked birds with bow and arrow, collected clams in the tidal pools along the Hudson, and stole chickens for "surreptitious roastings at fires in the woods." There were swimming holes in ponds and on the sandy beaches of the river. Boat docks provided a too-tempting platform for diving into the water, causing complaints from train passengers about naked children (Grinnell and his friends) "dancing on top of the piles, and generally making an exhibition." An ordinance was passed requiring swimmers to wear "tights," and when Grinnell and a friend ignored it, a policeman hauled them off to jail. "An hour or two of this confinement gave us plenty of time to ponder on the sorrows of life." A judge eventually sent them home.[7]

Grinnell's own imagination provided ample fuel for his childhood play, but an additional source helped to animate his boyhood adventures. Though barely known today, one of the most popular novelists of the mid-nineteenth century was an Irishman named Thomas Mayne Reid. Reid came to America in search of adventure in the 1840s. He found it, among other places, by enlisting with the U.S. Army during the war with Mexico. He earned a commission and fought in battles including Vera Cruz and Chapultepec, where he was seriously wounded. After the war, Reid began writing novels, most based loosely on his time in the American Southwest. He used the pen name Captain Mayne Reid, and his books, with titles such as *The Rifle Rangers*, *The Scalp Hunters*, and *The War Trail*, struck with particular resonance among young boys—and certainly with Grinnell. "His stories had appealed to my imagination." Indeed many of the heroes in Reid's books were boys, as in *The Young Voyageurs*, with the subtitle *The Boy Hunters in the North*, and *The Boy Hunters*, subtitled *Adventures in Search of a White Buffalo*.[8]

There were no buffalo of any shade near Audubon Park, but Grinnell found many opportunities to act out the scenes he devoured from the books of Captain Reid. "It must have been 1860, or possibly 1861, when I was eleven or twelve years old, that I first began to go shooting." Knowing that his parents would disapprove, Grinnell and a friend began secretly borrowing an old military musket from the village tailor. The gun was far taller than either boy and was so heavy that "neither could hold it to the shoulder" by himself. They took turns firing by having one boy, the shooter, rest the gun on the other's shoulder. No creature was safe. "Small birds were the chief game pursued ... meadowlarks, robins, golden winged woodpeckers and occasionally a wild pigeon." Later one of Grinnell's uncles legitimized the hunting activity when he gave Grinnell a small shotgun of his own.[9]

In Grinnell's writings about his childhood, no topic receives more attention than hunting. Indeed hunting became Grinnell's most important early connection to the wild world, teaching him skills in the close observation of nature. Hunting also provided a new tie to the Audubon family. One of Grinnell's frequent field companions was Jack Audubon, grandson of the painter. The connection went deeper still. Jack, in Grinnell's words, "was privileged" with carrying his grandfather's rifle—the same weapon that John James Audubon had relied upon during his epic adventures across the Mississippi.[10]

Audubon Park steeped young George Bird Grinnell in other icons of the West, a "region which in those days seemed infinitely remote and romantic with its tales of trappers, trading posts and Indians." The houses of Audubon's two sons were like museums, their walls adorned with trophies of deer and elk, powder horns, and ball pouches. A portrait of the buckskin-clad artist stared down, along with dozens of his paintings. In the barn, where the boys often played, were "great stacks of the old red, muslin-bound ornithological biographies and boxes of bird skins collected by the naturalist, and coming from we knew not where." Because the artist's sons carried on his work, they corresponded with scientists from around the world, and "frequently received boxes of fresh specimens." Grinnell remembered crowding around newly arrived shipments with his friends, waiting breathlessly for the cartons to be opened, then gazing in wonder at the "strange animals that were revealed."[11]

Of all the members of the Audubon family, none would exert greater influence on the young Grinnell than Lucy, the septuagenarian widow of the artist. Grinnell described her as "a beautiful white-haired old lady with extraordinary poise and dignity; most kindly and patient and affectionate, but a strict disciplinarian and one of whom all the children stood in awe." From her bedroom she ran a school for the children of the neighborhood. Some of them were her grandchildren; all of them called her "Grandma Audubon."[12]

Lucy Audubon appears to have been one of those unique teachers—a lucky person might encounter one or two—who changes the lives of their students. She understood the children she taught, grasping the significance of those fleeting moments of connection when an edu-

John James Audubon: Grinnell spent his childhood roaming the grounds of
the (then rural) New York estate of the former painter-naturalist.
Courtesy of the National Park Service.

cator has an opportunity to impress a young mind. "If I can hold the
mind of a child to a subject for five minutes," she said, "he will never
forget what I teach him."[13] George Bird Grinnell, certainly, would always
remember.

Some of the lessons were mundane: reading, writing, and 'rithmetic.
Others were more unique: Lucy Audubon had worked closely with her
husband on some of his later writings, including his *Ornithological Biography*. The experience had given her a solid exposure to natural science,
a knowledge she passed on to her pupils. This knowledge—and Grinnell's close relationship with his teacher—were on display in a story
Grinnell later recounted about a time he captured a live bird in a net. "I
rushed into the house and up to Grandma's room, and showed her my
prize. She told me that the bird was a Red Crossbill—a young one—
pointed out the peculiarities of the bill, told me something about the
bird's life, and later showed me a picture of it. Then after a little talk she

and I went downstairs and out of doors, found the birds still feeding there, and set the captive free."[14]

Some of the lessons that Grinnell learned from Lucy Audubon—including the one that would later emerge as the central creed of his life—would not register for years. Indeed from the time Grinnell left the school of Grandma Audubon to the time he graduated from college, his life had the feeling of casual indifference and missed opportunity.

His wealthy parents sent him to the best secondary schools. In 1861, when George was 12, he began a two-year stint at Manhattan's French Institute. At age 14 he enrolled in the prestigious Churchill Military School at Sing Sing. Churchill enforced a mild form of military discipline, but the accommodations were hardly spartan. The supplies that new students were instructed to bring to school included "napkin ring, bathrobe and slippers, mackintosh and umbrella, sponge and nail file." For three years, Grinnell went obediently through the well-defined motions of the school. He eventually rose up the ranks to command a company of his fellow students, but his academic performance was middling at best. Grinnell's self-appraisal was both insightful and blunt: "I knew very well that I had wasted my time at school."[15]

Nor did Grinnell have any sense of direction, a passive actor in setting the course of his life. "It had been determined that, when I left school, I should go to Yale, where my grandfather had graduated in 1804, and others of my ancestors had associations." Even with family connections, Grinnell's academic credentials made admission to Yale a dubious proposition. Grinnell's instructors at Churchill warned him that he was not prepared to pass Yale's rigorous entrance exams. But "my parents had made up their minds, and I was not in the habit of questioning my father's decisions." Grinnell spent the summer of 1866 in tedious remedial review. In September he traveled to New Haven and just managed to gain entrance, though "I had conditions in Greek and in Euclid."[16]

Having successfully put his nose to the grindstone to win admittance, Grinnell found that his lackadaisical attitude toward his education quickly resurfaced once on campus. "Little of interest happened" was his summary of his freshman year. Grinnell's sophomore year was more interesting because, as he explained, "I was perpetually in trouble." He did find application for his outdoor skills, climbing up the lightning rod of a campus clock tower in order to inscribe his class number at the top. Grinnell was also an enthusiastic participant in "all the hazing and hat-stealing which was usual by Sophomores." Partway through the fall semester, Grinnell was "detected in hazing a Freshman, and was suspended for one year."

Grinnell, along with a few fellow transgressors, was exiled to Farmington, Connecticut. Their supervisor in Farmington, one Reverend L.R. Payne (Yale '59), was responsible for tutoring the boys and, presumably, guiding them back to a more responsible path. Contrition, however, was in short supply among Grinnell and his comrades. "At Farmington we had a very good time, doing very little studying, and spending most of our time out of doors." It was almost like being back at Audubon Park. "We took long walks, paddled on the Farmington River, and on moonlight nights in winter used to spend pretty much all night tramping over the fields." At the end of the school year, Yale gave Grinnell the opportunity to take the exams with his class, but "[m]y idleness at Farmington resulted in a failure to pass."

The following autumn, no doubt at the intervention of his parents, Grinnell was set up with a new tutor, a physician named Dr. Hurlburt, this time in Stamford. Dr. Hurlburt, according to Grinnell, was "not only a good tutor, but a good handler of boys." His prescription for the wayward Grinnell: a "course of sprouts." Hurlburt roused Grinnell at an early hour, taking him along on his rounds. As they drove along in his buggy, Grinnell recited his lessons, then studied on his own while the doctor made house calls. At the end of two semesters, Grinnell returned again to Yale, this time passing his exams "with flying colors."[17]

Though back on track, Grinnell evinced no particular interest in where that track might lead. Inertia seemed to be the main force to compel him, and inertia seemed to be leading him toward a rather unremarkable life as a member of an upper-class, East Coast family. The summer before his senior year, his mother took him and two brothers on a three-month tour of Europe. The Continent appeared to make no particular impression; in his memoirs he recounted no detail of the trip beyond a list of the countries visited. On returning to Yale for his senior year, Grinnell was elected—in good Yale fashion—to Scroll and Key, a secret society. He offered no hint of what he might want to do after graduation, though he did describe himself as being "in great fear lest my degree should be withheld on account of my poor scholarship." Beyond this concern, his descriptions of his last year in college are bland: "I roomed in the old south college, and the room faced the green."[18]

Opportunity, though, was on the horizon—opportunity that would extend Grinnell's horizons far beyond the south college green.

IN THE SPRING OF 1870, A FEW MONTHS BEFORE GEORGE BIRD GRINNELL would graduate from Yale, a rumor swept the campus. A professor named Othniel Charles Marsh, it was reported, intended that summer to lead a scientific expedition to the far west. The expedition was to be manned by a dozen recent Yale graduates with the purpose of searching for dinosaur bones.[19]

For perhaps the first time in his life, Grinnell felt the spark of genuine inspiration. "This rumor greatly interested me," he remembered, "for I had been brought up, so to speak, on the writings of Captain Mayne Reid." Captain Reid, in fact, might almost have been speaking for Grinnell when he wrote in his novel *Wild Life* that "I as well as others yearned for the life beyond the confines of our secluded valley, and sighed for a participation in those deeds of which now and then a rumor reached our ever-attentive ears." Grinnell had long harbored the desire to visit the western scenes described by Reid and Audubon "but had supposed they were beyond my reach." Now, suddenly, an opportunity stood within range. "I determined that I must try."

It took several days for Grinnell to summon the courage to present himself before the intimidating Professor Marsh, who looked a bit like Ulysses S. Grant, the man then president. When Grinnell finally interviewed, the professor discouraged him, promising only to inquire about the young man's credentials. Grinnell, painfully aware of his record at Yale, could hardly have been optimistic. But when Grinnell went back for a second interview, he learned that he had been accepted as a member of the Marsh expedition.

For the first time, George Bird Grinnell was aiming in a direction that he himself had set—west. In a matter of weeks, Grinnell would depart on a five-month, 6,000-mile journey that would change the course of his life.

PROFESSOR OTHNIEL CHARLES MARSH WAS ABOUT TO EMERGE AS one of the most important scientists of his day. He was the nephew of George Peabody, a wealthy investment banker and a man considered by some to be the father of modern philanthropy. Today, museums throughout the Northeast still carry his name. Peabody was generous to Marsh too, having supported his education, first at Andover, then Yale, then in Europe. Marsh proved a brilliant student, discovering two important reptile fossils while still in college.[20]

During his time in Europe, Marsh convinced his uncle to give money for the establishment of a new natural history museum at Yale as well as an endowed chair in paleontology—the first in the United States. In 1866, Marsh returned to New Haven to occupy the endowed chair and to manage the museum. Two years later, Marsh made the discovery that set the stage for Grinnell's first great adventure.

In 1868, while traveling in the West, Marsh read an intriguing story in an Omaha newspaper. According to the report, a railroad construction worker had unearthed ancient bones while digging a well. They were, declared the paper, from the skeleton of an ancient man. Marsh was determined to investigate and managed to convince a train conductor to make an unscheduled stop at the site, a remote location known as Antelope Station. While the train cooled its engine, Marsh sifted through the mound of dirt beside the well. "I soon found many fragments and a number of entire bones, not of a man, but of horses diminutive indeed, but true equine ancestors." He had discovered *Protohippus*, a three-toed, miniature horse. "I could only wonder," he later wrote, "if such scientific truths as I had now obtained were concealed in a single well, what untold treasures must there be in the whole Rocky Mountain region?"[21]

Marsh intended to find out. He had attempted to line up an expedition for 1869, but intense Indian fighting prevented it. Finally, in the summer of 1870, the way was clear.

ADVENTURE FOUND GRINNELL ALMOST IMMEDIATELY AFTER THE arrival of the Marsh expedition at Fort McPherson, Nebraska, sixteen miles east from the confluence of the north and south forks of the Platte. Professor Marsh, through a personal relationship with General Philip H. Sheridan, had arranged for close cooperation on the part of the U.S. Army. The expedition visited the fort to rendezvous with its cavalry accompaniment before heading into the field.

On the day the Marsh expedition arrived at the outpost, a dozen Sioux Indians attacked a party of antelope hunters near the fort. One young Sioux warrior galloped up close before loosing an arrow, striking one of the hunters in the arm. The hunter in turn shot the Sioux, though the wounded Indian managed to ride off. Back at Fort McPherson, a troop of cavalry was dispatched to give chase. The soldiers took with them the post scout, William F. Cody, better known as Buffalo Bill. The troops failed to find the attackers with the exception of the boy who had been shot. He was found dead, wrapped in a buffalo robe on the top of a hill. Cody carried back the boy's moccasins and a few trinkets taken from the body. Grinnell remembered how "the newcomers from New Haven stared in wonder."[22]

Buffalo Bill himself, whom a breathless Grinnell described in his journal as "the most celebrated prairie man alive," led the Marsh expedition on its first day out of the fort. Professor Marsh had selected as their destination the Loup River, a desolate region of low, sandy hills in the heart of Sioux territory. The cavalrymen protecting the expedition were commanded by Major Frank North, an officer with a national reputation as a famous Indian fighter. Two Pawnee—bitter enemies of the Sioux and the Cheyenne—scouted ahead of the column, sometimes crawling on their bellies through the grass to peer over hilltops before advancing. Grinnell and the other civilians rode on Indian ponies, captured from the Cheyenne at the Battle of Summit Springs. Six army wagons brought up the rear with food, ammunition, and tents.[23]

On their first night in the field, Professor Marsh delivered a campfire talk to an assembly that included Cody and most of the cavalrymen. For his topic, Marsh selected . . . geology. The lecture, like the broader purpose of the Marsh expedition, "greatly puzzled our military companions of the rank and file." Nevertheless, Marsh attempted to enlighten them with explanations of various rock formations, how they had come to be formed, and the discoveries of ancient beasts that he hoped lay ahead.

Buffalo Bill, though described by Grinnell as an "interested auditor," remained among the skeptics, remarking afterward that "the professor told the boys some mighty tough yarns today."[24]

After two or three days of hot, waterless marches, Grinnell concluded that he and his companions "had seen quite enough of Nebraska." Certainly they understood why the literature of the time referred to the plains as the Great American Desert. On one stop, they measured the temperature at 110 degrees Fahrenheit in the shade. What little surface water they found was impregnated with alkali, so that even the animals could not drink it. To obtain potable water, they had to dig in the bed of dry lakes.[25]

Conditions improved somewhat when they hit the Loup River, where water was available and game was more plentiful. Members of the expedition saw their first elk, and each day Major North allowed one of the young explorers to accompany him and to shoot at antelope. No member of the Marsh expedition managed to bring down a pronghorn, the fastest land animal in North America. Major North and the Pawnee scouts, fortunately, had better luck, and the party was well fed.[26] The presence of water, as Grinnell noted in his journal, did come with its own set of problems. "I was delighted by discovering what I supposed was a new species of heron," he cracked, "but upon examination I found it was only one of the mosquitoes of the country.[27]

Grinnell and his companions improved their firearms proficiency as their voyage continued. Rattlesnakes provided frequent targets, "for the country swarmed with the reptiles." The snakes bit three of their horses. "Their humming soon became an old tune; and the charm of shooting the wretches wore away for all but one, who was collecting their rattles as a necklace for his lady-love."[28]

At the Loup, the expedition finally began its excavations, digging into ground called *mauvaises terres*—the bare clay of ancient lake beds where fossils were most likely to be found. The cavalry stood guard while Marsh and his assistants dug, ever mindful of their dangerous intrusion into Sioux land. The early findings of the party lived up to Professor Marsh's expectations: camels, ancient horses, a mastodon, and some species new to science. On several occasions the expedition encountered Indian burial grounds, the bodies set up on high platforms. Beneath the platforms were skeletons of horses, "killed to provide the owner with mounts in his future life." Though fearful of smallpox, the party was not above adding a few of the Indian skulls to their collection.[29]

The nearest sign of live Indians came one night as they settled in along the Loup. A prairie fire erupted simultaneously on both sides of the camp—set, they believed, by the Sioux. The soldiers quickly set backfires to rob the advancing flames of fuel. "When our anxiety with regard to the camp had subsided," wrote Grinnell, the fire was "very beautiful." For two days, though, the expedition would march over burnt grass, struggling to find forage for their horses.[30]

After ascending the Loup River to its headwaters, the expedition turned south and west for the second stage of their exploration—the triangle of land between the north and south forks of the Platte.

As the Marsh expedition hunted for dinosaur bones along the Nebraska–Wyoming border, they followed a trail that was itself an artifact, though of far more recent vintage. Grinnell referred to it as the "old California and Oregon trail." It had been only fifteen months since the driving of the golden spike had connected the continent by rail, but the age of transcontinental travel by wagon was already history. The deep ruts that so recently conveyed hundreds of thousands of emigrants to the West were now returning to prairie. "The grass was growing all along the road," wrote Grinnell, "and where wagons had passed was a continuous bed of sunflowers."[31]

The cavalrymen who accompanied the explorers provided a constant reminder that the territory they crossed remained contested—particularly an area along a North Platte tributary called Horse Creek. "This is famous hunting ground," wrote Bill Betts, one of Grinnell's young colleagues, "and we came upon many fresh signs of savages." For Grinnell, though, the creek offered an enticement that he could not resist. "Notwithstanding these evidences of unfriendly neighbors," continued Betts, "two of the party, all intent on duck-shooting, persisted in following the creek."[32]

Armed only with shotguns, Grinnell and a friend named Jack Nicholson split away from the main party and headed up Horse Creek. The rest of the expedition, meanwhile, took a shortcut—away from the creek and across the open prairie. A cavalry captain named Montgomery told

Grinnell and Nicholson that if they followed the creek, they would eventually reach the place where the main party would camp for the night. The plan was to meet up before sundown, hopefully with enough ducks for the entire mess. What Captain Montgomery did not know, what no one knew in this unfamiliar country, was that Horse Creek took a lengthy diversion before arriving at the place where the main party intended to camp. To reach the campsite by following the creek would have required a trek of at least fifty miles—more than twice the distance that Grinnell and Nicholson had been told to expect.

The hunting, at least, went according to plan. Ducks abounded, and if nothing else, the two men would be well fed. All day long, the hunters kept their horses along the creek until finally, with sunset nearing, they began to become concerned. Surely they'd come more than twenty miles. After taking a break to smoke and to mull over the situation, they decided to ride up a nearby butte for a better view of the country.

From the vantage of the hill, Grinnell and Nicholson learned two things, neither good. Despite being able to see for miles, they could find no sign of the Marsh expedition's encampment. Clearly they would not reach their companions by nightfall. The second negative development came from behind them, from down the hillside at the very place where they had just talked and smoked. After lighting their pipes, one of the men (it's not clear who) had thrown his match into the tall grass. A "roaring prairie fire" now raced up the hill toward them.[33]

CHAPTER TWO

"Self-Denial"

*He triumphed in the strength of another, who molded his
character, shaped his aims, gave substance to his dreams,
and finally, by the exercise of that self-denial which he was
incapable of as a long-sustained effort, won for him the
public recognition and reward of his splendid talents.*

—George Bird Grinnell
on John and Lucy Audubon

G rinnell and Nicholson's initial instinct was to spur their horses
and run, but the wildfire—whipped by a vicious wind—
burned up the hill "at an inconceivably rapid rate." A backfire,
they realized, was their best defense. As they had seen the soldiers do on
the Loup River, they quickly set a small blaze. The backfire, as Grinnell
described it, "checked the flames so that we were able to reach a gravelly knoll." There, they held on to their horses as the wall of flames
burned around them, "near enough to singe the hair on our faces and
on the horses."[1]

As they caught their breath and soothed their mounts in the wake of
the blaze, the lost hunters felt a renewed sense of urgency about hostile
Indians. The smoke from the fire, they feared, would draw the attention
of anyone within a range of several miles. Armed only with shotguns,
meanwhile, the two hunters would have little ability to stand off a hostile approach. They briefly considered cutting across the open plains in
the hope of picking up the expedition's trail but quickly decided against

Marsh expedition in 1870, near Fort Bridger in what is today Wyoming.
Professor Othniel Marsh is the bearded man in the center. Grinnell is
third from the left in the broad-brimmed hat.
Courtesy of the Scott Meyer Family.

such a course. Better to backtrack, sticking close to Horse Creek—a
"guide that would not fail them." Once back to their starting point, they
could easily find the tracks of Marsh, the cavalry, and the rest of their
party. In the meantime, night was nearly upon them.

Fearful of attack, Grinnell and Nicholson resorted to ploy. "As the
sun was setting we stopped on the borders of the stream, unsaddled our
horses and built a fire, to convey the impression that we had gone into
camp." They plucked and roasted some of their ducks, filling their bellies while waiting for the full cover of darkness. Then, saddling up, they
rode into the creek. For a long time they rode down the little stream,
using the water to cover their tracks. Only when they'd covered a mile
or more did they climb up onto the bank, there to spend a long, frigid
night with no fire.

By the next morning, Professor Marsh and the cavalry accompaniment
were convinced that the two missing hunters had fallen victim to the Cheyenne, the most pervasive tribe in that region. The concern seemed to find

confirmation when a search party came across two lame Indian ponies. The Cheyenne, it appeared, had killed Grinnell and Nicholson, taken their two horses, then left the crippled stock behind. But by the time the search party made it back to camp, Grinnell and Nicholson had reappeared, having backtracked and then followed the expedition's clear trail.

Grinnell was nonchalant about the entire incident, his main comments reflecting pride that their evasive tactics on Horse Creek had kept the cavalry from picking up their trail. "The devices we had practiced to mislead the Indians were so successful that on the following day the searchers lost our trail hopelessly." Certainly his remarks reflected the young-man bravado that puffed the chests of many who traveled the wild lands of the West. But something more profound was also at play. Even as he labored each day as a member of an expedition whose findings would stun the scientific world, Grinnell's own journey was more personal. On an expedition whose aim was discovery, Grinnell was discovering himself.

If there were two moral poles in the world of George Bird Grinnell, Cornelius Vanderbilt stood at one of them. Vanderbilt was the most important client of the brokerage house owned by Grinnell's father, and Grinnell had known him from an early age. Commodore Vanderbilt (the title reflected ownership of a veritable flotilla, not a military background) stood at the helm of the era that would rise in the aftermath of the Civil War—the era in which Grinnell would spend the most important years of his life. Mark Twain, in his biting 1873 satire, dubbed it the "Gilded Age." Beginning in the boom years that followed the Civil War and extending to the presidency of Theodore Roosevelt, the Gilded Age was the heyday of the robber barons, a class of industrialists known as much for their plundering business style as for their towering business achievements.

A survey of Vanderbilt's life reads like a primer for the whole ilk. He offered a glimpse of what would prove to be his modus operandi in his very first enterprise. In 1811, at the age of 17, Vanderbilt borrowed money from his mother to start a ferry in New York Harbor. With screaming Oedipal overtones, he hoped to capsize the venture of his main rival—his father.

At every juncture in his long career, Vanderbilt was in the right place at the right time with the right service—transportation. His wealth grew during the War of 1812, the 1849 gold rush, and then the Civil War. After dominating New York ferry traffic, he later came to control coastal trade between New York and New England. When George Bird Grinnell was a boy, Vanderbilt was selling his shipping enterprises and buying railroads. By 1869 he would own the New York Central and Hudson River Railroad, controlling much of the lucrative transport between New York City and the Great Lakes.

Business for Vanderbilt was war. He once wrote to an enemy promising that "I won't sue you, for the law is too slow. I will ruin you." Such was Vanderbilt's reputation that rivals in a shipping venture to Nicaragua paid him $50,000 a month for his promise to stay out of the market.

Vanderbilt, in short, embodied the characteristics of the Gilded Age in which Grinnell lived. As a Vanderbilt biographer described him, he was "simply the most conspicuous and terrifying exponent of his era," an era "of ruthlessness, of personal selfishness, of corruption, of disregard of private rights, of contempt for law and Legislatures, and yet of vast and beneficial achievement." Vanderbilt would die with a fortune estimated at $100 million, though not before cutting several sons out of his will.[2]

Securing financing for Vanderbilt's railroads would be the culminating professional achievement of Grinnell's father, and for a while, Grinnell appeared likely to follow in his father's footsteps.

It was Lucy Audubon, ultimately, who would put him on a different track. She taught him a philosophy startling in the degree to which it ran counter to the prevailing ethics of the day, a discipline that leading men of Grinnell's generation were more likely to scorn than to follow. It was a value system that stemmed from the central narrative of her life.

Born Lucy Bakewell in England in 1787, she was the daughter of a country gentleman of modest wealth. Lucy received top-flight schooling for a girl of her generation. Her father believed that girls should be educated, if only to make them better and more engaging wives. Lucy was tutored from a young age and later attended boarding school, where she studied French, dance, and needlepoint. Her formal education was supplemented by tutors and long hours of reading in her father's extensive library. Her father also taught her to ride, and Lucy loved the outdoors from an early age.[3]

Lucy Audubon: The creed
of Grinnell's boyhood tutor,
"self-denial," flew in the face
of the conspicuous consumption
of Gilded Age America.
*Courtesy of the New York
Historical Society.*

In 1801, when Lucy was 14, she moved with her family to America, eventually settling in a Pennsylvania country estate with the bounteous name of Fatland Ford. They lived a half mile from the estate of a well-to-do French family, the Audubons. Overcoming a residue of British–French antipathy, Lucy's father became a hunting partner of John James, the Audubon's dashing, long-haired son. When the young Audubon first met Lucy, he recorded in his diary that he was smitten by her "*je ne sais quoi*" (an attribute not so readily apparent in her photos as the quintessential old widow). The two married in 1808 and moved to Kentucky, where John James began a decade-long series of financial misadventures. By 1819, the Audubons were bankrupt, and for a brief time, John James was even jailed for his debts. There would be other, even greater tragedies—the Audubons lost two infant daughters to sickness. Two sons, Gifford and Woodhouse, would survive childhood—though they would not outlive their mother.[4]

Having failed at business, Audubon was determined to make a living in the field that fueled his passion—painting. He scraped by—drawing portraits at $5 a head, teaching art at two girls' schools, and working as a taxidermist in a museum. Since a young age, though, Audubon had loved to paint birds, and it was toward this endeavor that he increasingly turned. He began to travel in search of new subjects, eventually building a portfolio detailing hundreds of species. But it would be decades before he achieved any kind of financial reprieve. In the meantime, John James was often away from home for long periods of time, sometimes years, as he traveled the American West and later toured Europe in an effort to secure patrons and a publisher.[5]

Lucy Audubon, meanwhile, who had grown up in a life of privilege and wealth, became the breadwinner for herself and her two young sons. One of the few professions open to women was teaching, and Lucy began to conduct classes out of her home. Later she and her children took up residence with a wealthy plantation owner in Louisiana, teaching the local children in exchange for a place to live and a small salary. For two decades, she faced constant fears about putting food on the table, the indignities of asking for credit, and the struggle of single parenthood. It was a time of remarkable sacrifice—of subordinating her own needs and desires in favor of her family. Indeed self-sacrifice became the creed by which she lived, a way of putting her life into a broader context. While Audubon struggled for recognition and financial success, Lucy supported him, encouraged him, guided him, and advised him with a sound judgment that the artist lacked. And she held the family together.

For John James Audubon, recognition would come long before financial success. With Lucy's encouragement, Audubon in 1826 took his collection to England. In Europe, Audubon would earn fame as both an artist and a naturalist. But not until the publication of *Birds of America* in the late 1830s did the Audubon family achieve its first semblance of financial security. In 1841, when Lucy was 54 years old, Audubon bought the tract of land along the Hudson River that would become Audubon Park. In 1842 he completed construction of a grand house, proud finally to have provided the home that he felt his wife deserved. He called it Minniesland, a reference to his pet name for Lucy.

For a few years at Minniesland, Lucy enjoyed a short interlude of repose. Her decades of hard work, it appeared, had come to fruition. With his two sons assisting him (and living in their own houses at Audubon Park), John James labored on the last of his great works, *Viviparous*

Quadrupeds of North America. Lucy took pleasure in her home, the company of her sons, and a growing brood of grandchildren. But it would not last.

In 1844, Audubon's eyesight began to fail him. Soon he could no longer paint the exacting details that defined his work. In 1847, Audubon suffered a stroke that crippled his mind. He spent the last years of his life in a desperate fog of senility, dying in 1851 at the age of 65. Lucy had known tragedy before, but still worse events lay ahead. In 1856, her son Gifford suffered a fall that left him an invalid. Lucy rented out the house she loved and moved in with Gifford and his family, helping to care for him.

In 1857—the same year that George Bird Grinnell and his family moved to Audubon Park—Lucy began teaching again to supplement the family income. Grinnell would later write that Grandma Audubon "seemed to be doing for her sons and their families something like what she had been doing for her husband during much of the time of their marriage, earning the bread for the family."[6] She began to sell off pieces of Audubon Park, including the property purchased by Grinnell's father. Young George accompanied his father to the closing, and he was struck by Grandma Audubon's "great relief, satisfaction, and even gratitude." The scene moved him, "though for years afterward I did not understand its meaning."[7] Even with her land, Lucy's assets could provide no protection against the next waves of calamity to strike her family.

The sons of John James Audubon apparently inherited their father's weakness in matters financial. In Grinnell's words, "neither he nor his sons were businessmen." In 1859, Woodhouse Audubon invested a large sum in the publication of a new edition of his father's *Birds of America.* The new book was sold to subscribers, most of whom happened to reside in the South. When the Civil War broke out, the investment turned into a near complete loss. Creditors placed liens on Audubon Park.

Gifford, meanwhile, having languished since his accident, died in 1860, Lucy at his side. Lucy would soon watch the death of Woodhouse too. Devastated by the death of his brother and the pressure of his financial losses, Woodhouse fell ill. He died in 1862. The next year, to stave off her sons' creditors, Lucy sold her remaining property at Audubon Park and moved in with a granddaughter in Washington Heights. Over the years, she sold off paintings from her husband's collection, including the original plates for *Birds of America*, which at least provided an income until her death in Louisville at the age of 87.

One painting that Lucy did not sell was a large work by her husband called *The Eagle and the Lamb.* The painting must have had special significance: She hung it, out of all her husband's works, in her bedroom. The painting, as she knew, was also a favorite of her student George Bird Grinnell. Before her journey to Louisville, the last journey of her life, Lucy wrote a short note in a shaking hand to Grinnell. "Dear young friend," it began. She worried in the note about the possibility of an "accident" on the trip, perhaps her way of expressing concern about the hardship of travel on the very old. If anything were to happen, she directed Grinnell to "take possession of the Eagle & Lamb, with all the love & esteem for yourself & parents that is possible for our hearts to feel." Lucy survived the trip to Louisville, though she died shortly thereafter. Her will would also specify that the painting should go to Grinnell. It hung in his home for all his life.[8]

Grinnell would sum up the life of Lucy Audubon in an essay he would later write about her husband: "The great lesson of his life lies in our recognition . . . that he triumphed in the strength of another, who molded his character, shaped his aims, gave substance to his dreams, and finally, by the exercise of that self-denial which he was incapable of as a long-sustained effort, won for him the public recognition and reward of his splendid talents."[9]

The greatest lesson that Grinnell would learn from Lucy Audubon—the greatest lesson of his life—was her creed of self-denial. Self-denial, preached Grandma Audubon, was the "key to success in life."[10] For Lucy, sacrifice made possible the success of her husband and the stable upbringing of her children. Their successes became hers, and appropriately so.

Grinnell, only fourteen at the time that Lucy left Audubon Park, could not yet know the importance of the boyhood lesson that he would one day apply to the broad world around him. His teacher had sown a seed, but the conditions for cultivation were not yet present. For many years, in fact, it appeared that young George Bird Grinnell might instead follow the path of those to whom too much is given. In the West with the Marsh expedition, though, came the first signs of Grinnell's great awakening.

THE BASE OF OPERATIONS FOR THE THIRD AND MOST SIGNIFICANT stage of the 1870 Marsh expedition was the southwest Wyoming outpost of Fort Bridger. Jim Bridger, "King of the Mountain Men," had established the fort in 1843, and it became an important way station for westbound travelers. Now two months into their voyage, Grinnell was proud that he and his Yale companions had come to look like the frontiersmen they idolized: "Bearded, bronzed by exposure to all weather, and clothed in buckskin, you might take them at first glance for a party of trappers."[11]

For the Marsh expedition, Fort Bridger was the gateway to the greatest scientific discoveries of the journey. In the basin to the south of the fort, the expedition uncovered *Eohippus*, the earliest known member of the horse family, and also the "extraordinary six-horned beasts later described by Marsh as *Dinocerata*."[12]

Grinnell certainly shared in the pride and excitement of these discoveries. In his memoirs he noted briefly that the "three trips that the expedition had already made resulted in great collections of fossils, which were of extraordinary interest."[13] Grinnell had earned the respect of Professor Marsh, known as a difficult taskmaster. Indeed the 1870 expedition would represent the beginning of a long professional relationship between the two men. Still, Grinnell's writings leave little doubt about the aspect of their journey that excited him the most, and it was not paleontology.

Grinnell would be a prolific writer throughout his life, including two dozen books, hundreds of magazine and journal articles, thousands of letters, and a short, unpublished memoir. While his writing could be forceful and passionate, particularly when advocating for a cause in which he believed, it rarely offered much insight into Grinnell as a person. (This is particularly notable in contrast to today's "reality" tidal wave of shallow introspection ad nauseum.) Yet in his writings about the land surrounding Fort Bridger, Grinnell offered several revealing glimpses of the degree to which he was moved by what he saw around him. "North, south, east and west the eye rests upon mountains piled on mountains," he wrote of the Uinta Range. "Truly it is a grand scene, and a lover of nature may well be exercised if, for a time, he forgets all else in contemplating it."[14] Of the Green River, Grinnell wrote that "[e]ach mile of the river's length presents fresh charms, and the thoughtful mind is awed and purified by the contemplation of these, some of the grandest works of Nature."[15]

Grinnell, who had come to the West in search of the "wild and wooly," had found something more profound. At times, Grinnell's descriptions took on the explicit overtones of religion:

> Parks there are, where the tall pines and the cottonwoods, with their silvery foliage, stand as if arrayed at the command of the most skillful of gardener; where green meadows, dotted with clumps of trees, or with little copses, stretch away toward the rocky heights beyond and seem almost to reveal the hand of man in the artistic beauty of their design. But no gardener planted these towering trees, nor was human skill evoked to lay out these delightful parks; the hand of a greater being than man is visible in all these beauties—the hand of God.[16]

On September 20, 1870, George Bird Grinnell celebrated his twenty-first birthday in a ruggedly beautiful campsite along the Henry's Fork River. After a cheerfully aimless youth, Grinnell was entering adulthood with bigger thoughts on his mind.

In addition to a view of nature in full pristine splendor, the Fort Bridger stage of the Marsh expedition also exposed Grinnell to "a glimpse of the old-time trapper's life in the Rocky Mountains of thirty years before." Grinnell needed a new horse and was told that three trappers encamped on the Henry's Fork might be willing to trade. He rode out to find Ike Edwards, Phil Mass, and John Baker. Each trapper had an Indian wife and "a large flock of children of various sizes." The oldest of the trappers had been in the West for decades and experienced the height of the beaver-trade era. Grinnell was fascinated by the fact that "they still supported themselves, in part at least, by trapping beaver." They invited Grinnell to stay over. He eagerly accepted and "spent some days full of joy and interest in this old-time camp."[17]

They slept in teepees and lived off the land. "The river bottom and the hills were full of game; the stream full of trout." In the early mornings, Grinnell accompanied the old mountain men on forays to run traplines, usually returning with "one, two or more beaver." It was a lifestyle that Grinnell embraced to his core. "Their mode of life appealed strongly to a young man fond of the open," he remembered, "and while I was with them I could not imagine, nor can I imagine now, a more attractive—a happier—life than theirs." Nor was there any hint of exaggeration when Grinnell wrote that "I desired enormously to spend the rest of my life with these people."[18]

By October, though, the Fort Bridger stage of the Marsh expedition was complete. The party left behind the high adventure of the Rockies "and then spent several weeks in seeing what all tourists see." In Salt Lake City, Professor Marsh discussed his fossil findings with Brigham Young, who believed that the ancient horses provided evidence for the existence of the lost tribe of Mormon and Moroni described in the Book of Mormon. Marsh's young field assistants, meanwhile, "flirted with twenty-two daughters of Brigham Young in a box at the theatre." Then it was on to California, where the expedition visited Yosemite Valley and the "Big Trees." By Thanksgiving, Grinnell found himself back at Audubon Park, contemplating his future in the company of his parents.[19]

To a degree that Grinnell himself might not yet have appreciated, his adventures with the Marsh expedition represented the most important formative events of his life.

The expedition itself was a scientific triumph. Marsh carted thirty-five boxes of fossils back to Yale, where the contents became the foundation for the great Peabody Museum collection. In addition to *Eohippus*, Marsh and his assistants had discovered 100 extinct species new to science. Marsh, already a rising star, emerged from the voyage as one of the preeminent paleontologists of his day. His genealogy of the horse (eventually constructed using findings from this and subsequent Marsh expeditions) was acknowledged by no less than Charles Darwin as the most significant evidence for the theory of evolution since the 1859 publication of *Origin of Species*. Marsh's young assistants won their own share of fame, including a detailed article about their adventures in *Harper's New Monthly Magazine*, one of the most widely read journals of the day.[20]

The Marsh expedition provided Grinnell with a unique exposure to the West that would shape his thinking in fundamental ways. He became one of the last members of his generation—of any generation—to experience the West as it had been. Grinnell had traversed unexplored land, hunted for subsistence, dodged hostile Indians, and for a brief time lived the life of an 1830s trapper—anachronistic even in 1870.[21]

Yet even as he lived out this "primitive West," as he called it, Grinnell was on the cutting edge of something wholly new. Professor Marsh, in the words of historian William H. Goetzmann, "had discovered a new Western horizon." Marsh—with Grinnell along for the ride—was in the forward guard of a new generation of explorers. Their predecessors had seen the West primarily through the prism of territorial rivalry and/or commercial opportunity. Marsh saw buried in western sands the "rich rewards" of "scientific truths."[22]

There was another important aspect to Grinnell's time with the Marsh expedition. Five months of digging up fossil remains had given him a tangible experience with the fragility of life—indeed the fragility of whole species. Like Lucy Audubon's philosophy of self-denial, the notion that nature was fragile ran directly counter to a core tenet of Gilded Age and robber baron belief—the "myth of inexhaustibility."[23] According to this belief, nature, and the western United States in particular, was an untapped land of Goshen, a collection of unbounded resources to be developed and exploited for the betterment of humankind. A generation of early western settlers, and then their children in the Gilded Age, drew inspiration from writings such as Lansford W. Hastings's *The Emigrants' Guide to Oregon and California in 1844:*

> *I can not but believe, that the time is not distant, when*
> *those wild forests, trackless plains, untrodden valleys, and the*
> *unbounded ocean, will present one grand scene, of continuous*
> *improvements, universal enterprise, and unparalleled commerce:*
> *when those vast forests, shall have disappeared, before the hardy*
> *pioneer; those extensive plains, shall abound with innumerable*
> *herds, of domestic animals; those fertile valleys, shall groan*

under the immense weight of their abundant products; when
those numerous rivers shall teem with countless steam-boats,
steam-ships, ships, barques and brigs; when the entire country,
will be everywhere intersected, with turnpike roads, rail-roads
and canals; and when, all the vastly numerous, and rich
resources, of that now, almost unknown region, will be fully and
advantageously developed.[24]

Hastings's vision of complete human conquest provides a stark counterpoint to Grinnell's celebration of the wild Green River landscape created by "the hand of a greater being than man."

Grinnell's first reactions to the West were more visceral than intellectual—but they were not superficial. He craved the adventure and was deeply affected by the beauty of what he saw. In an 1873 article in *Forest and Stream* magazine, Grinnell would write of the Green River Valley that "the sportsman or naturalist will find here much to attract and delight him. And perhaps he will even be tempted, as I once was, to sever for a time the ties that bind him to his eastern home."[25]

Far from severing his ties to the East, though, Grinnell was about to secure them. In the contest between Lucy Audubon and Cornelius Vanderbilt, Vanderbilt appeared to have won.

Less than a month after his return to New York, remembered Grinnell, "I entered my father's office at 36 Broad Street, as a clerk without pay." His father, by 1870, was Vanderbilt's principal Wall Street agent. His father's business partner was Horace F. Clark, Vanderbilt's son-in-law. George Bird Grinnell, it appeared, would follow in his father's footsteps. "[O]n talking it over with my father I found that he was anxious to have me go into business, to relieve him, and ultimately to take his place. This seemed the proper thing to do."[26]

The Marsh expedition had awakened in Grinnell a deep passion, but for the time being anyway, he would cede to the wishes of his father. The great engine of the Gilded Age was barreling forward, and instead of jumping off, Grinnell was climbing aboard. As for pursuit of his western dreams, "the knowledge of the grief this would give my parents pulled me back again."[27]

CHAPTER THREE

"Barbarism Pure and Simple"

Here was barbarism pure and simple. Here was nature.

—GEORGE BIRD GRINNELL,
July 1872

I had always had a settled dislike for the business," was George Bird Grinnell's understated summary of his time at his father's firm, Geo. B. Grinnell and Company.[1] After the adventure and excitement of five months in the wild and wooly West, Grinnell settled in, as best he could, to his new life as a Wall Street broker. With his father as tutor, he learned the basics of buying and selling stocks. There were also more complicated lessons, including the high-stakes transactions of purchasing stocks—especially railroad stocks—"on a margin." The risks of such deals would later become painfully apparent. But in the exuberance of the early 1870s, Grinnell remembered that there was no thought but that prices "would go much higher."[2]

Grinnell, eager to please his parents, kept his nose to the grindstone during the long and tedious days at the brokerage house. His escape came in the evenings, when he retreated to Audubon Park, there to pursue his love for the outdoors through various vicarious activities. One outlet was a continuing relationship with Professor O.C. Marsh. Marsh had asked Grinnell to keep his eye out for the fossils and osteological materials that were "constantly coming into the menageries and taxidermists' shops in New York." When Grinnell found interesting pieces, he secured them for Marsh. Grinnell also took up a hobby popular among hunters and anglers of the day, taxidermy, with birds as his particular focus. Most nights, Grinnell wrote, he would spend "two or three hours of the evening down in the cellar, where I had an excellent outfit for mounting birds."[3]

None of these activities, though, sufficed to fill the void he felt for true adventure. "In the summer of 1872 I was anxious again to go out West." In particular, Grinnell wanted to undertake the quintessential western activity of the day—the buffalo hunt. "[T]hese hunts of the Indians [had] been described to me with a graphic eloquence that filled me with enthusiasm as I listened to the recital, and I had determined that if ever the opportunity offered I would take part in one." With the assistance of Major Frank North (who had guided the Marsh expedition up the Loup River), Grinnell arranged for a hunt, and not just any hunt. Grinnell, guided by Frank North's younger brother, Luther "Lute" North, would accompany an entire tribe of Pawnee Indians on their annual foray in the wild Republican River Valley of western Kansas.[4]

THE BUFFALO THAT GEORGE BIRD GRINNELL, AT AGE 22, SOUGHT TO kill, had been a fixture on the Great Plains of North American for thousands of generations.

Ancestors of the modern buffalo walked across the Bering Land Bridge from Siberia sometime between 300,000 and 600,000 years ago. One of these ancestors, *Bison latifrons,* was 20 percent larger than a modern buffalo and carried an intimidating rack of horns spreading seven feet. *Bison latifrons* shared the Great Plains with a number of the animals whose fossils Grinnell had dug up during the Marsh expedition, including miniature horses, mammoths, camels, and mastodons. These grazing animals were hunted by fierce predators, now extinct, including several species of saber-toothed tigers and the dire wolf, a larger version of its modern descendant.[5]

It is not clear exactly when humans first appeared on the Great Plains, though we do know that by at least 12,000 years ago, human hunters were among the predators stalking bison. We also know that around the same time, most of the large mammals of the plains became extinct. It is unknown whether they died because of human hunters, the climate change that followed the last ice age, or some combination of the two.

Whatever the cause, a few resilient survivors were able to adapt to their rapidly changing world. One was mankind. Another was the modern wolf. Another was a new species of bison that zoologists would one day name *Bison bison*—the modern buffalo. *Bison bison* had at least one advantage that helped it survive against primitive human hunters. More ancient species of buffalo, with their gigantic horns, defended themselves by standing and fighting. Armed with a long spear, a hunter could defeat this strategy. *Bison bison* had a different defense, one more difficult for man to overcome. Instead of fighting, *Bison bison* ran away.[6]

The modern buffalo, winner of a brutal contest that wiped out hundreds of other species, is a survivor of stunning physical attributes. Though its ultimate defense is to run away, the buffalo projects a physical presence that is intimidating to most predators. Bulls can weigh more than a ton, with thickly furred heads and leg pantaloons that make them look larger still. Both male and female buffalo have hooked horns, far smaller than those of their ancient ancestors but still plenty potent. With powerful neck muscles to support their massive heads, a buffalo can throw a wolf thirty feet. In breeding season, a male buffalo can kill a rival with one goring stab of his horns. Indeed about 5 percent of mature bulls die each year from wounds they receive in battle with their peers.[7]

Having survived both the Ice Age and its aftermath, buffalo can thrive in a remarkable range of climates, from 110 degrees Fahrenheit on the deserts of Mexico (or Nebraska) to −50 degrees Fahrenheit on the windswept plains of Canada. A buffalo's thick coat has ten times more hairs per square inch than the hide of a domestic cow. Yet in the summer, the buffalo sheds down to a thin coat as sleek as a lamb, newly shorn for the county fair.[8]

The buffalo and its relatives (including deer, elk, antelope, and the domestic cow) are ruminants, with digestive systems well attuned to subsist on the grasses of the Great Plains. Humans can't eat grass because we can't digest cellulose. Buffalo can because the first chamber of their alimentary canal—the rumen—is a sort of vat in which colonies of bacteria help to break down cellulose into usable carbohydrates. To further promote the process, ruminants chew their food twice: once before swallowing and a second time after fist-sized portions—the cud—are regurgitated and then chewed again.[9]

The buffalo's gigantic head serves a purpose beyond intimidation. In the winter, it becomes a powerful plow to push snow away from buried grass. Cattle, by contrast, have no such ability to dig for food. If not fed by ranchers during an extended period of heavy snow, cattle die.

Buffalo reproduce in prodigious numbers, at least by comparison with other large mammals. Buffalo cows drop their first calf at age 3, and in domesticated herds they have been bred beyond age 30. If a herd is well nourished, 85 to 90 percent of mature cows become pregnant and give birth each year. Farmers of modern dairy cattle, by contrast, using high-tech, artificial insemination, might achieve a birth rate of only 50 percent.[10]

Buffalo calves are born ready to run. Within two minutes of birth, a buffalo calf tries to raise itself. Within seven minutes, it is standing. Within an hour, a buffalo calf can run after the herd.[11] In a domesticated herd, a newborn calf was once herded sixty-six miles, through deep snow, in its first two days of life, with no apparent ill effects.[12] Such early skill and endurance is vital, since running—and the herd itself—are a buffalo's primary defenses.

Despite the buffalo's plodding, cumbersome appearance, it is shockingly fast and agile. An adult buffalo is as fleet as some racehorses in the quarter mile, but unlike a horse, a frightened buffalo can run across the prairie for miles without stopping. A nineteenth-century scientific expedition once required a relay of three fresh horses to run down a buffalo cow, covering twenty-five miles in the chase. As for agility, a biologist at the National Bison Range in Montana once observed a 2,000-pound bull leap up a six-foot embankment from a standing start![13]

The buffalo's skill at running away is enhanced by the collective power of the herd to tell it *when* to run away. The phrase "herd mentality" may carry a pejorative connotation for humans, but not so the buffalo: Hundreds (or thousands) of pairs of eyes, scanning the broad horizon. Hundreds of noses, sniffing for foreign scent. And then, at the first sign of danger, a powerful instinct to follow after the fleet, fleeing mass. When pursued, the herd protects its members through sheer numbers. A buffalo doesn't need to be faster than predators; it just needs to be faster than a few of the other buffalo.

Against the buffalo herd, even a predator as effective as the wolf represents little risk except to the young, the old, and the injured or sick.[14]

Healthy adult buffalo (not including cows with calves) pay little attention to wolves. Indeed Indian hunters sometimes cloaked themselves in
wolf skins in order to crawl within easy shooting distance. The power of
the herd (especially its size) even insulated the buffalo against its most
effective predator—humans.

The question of how many buffalo walked the North American
continent before the arrival of Europeans is a source of considerable
disagreement. Sixty million has been a common number cited, and
some estimates are as high as 100 million. The late Dale Lott, a biologist
at the University of California, Davis, and a leading expert on the buffalo, studied these higher numbers and made a persuasive case as to why
the number of 30 million—though still an educated guess—is the better
estimate based on the continent's carrying capacity.[15] Still, 30 million is
an enormous number.

Equally impressive is the range that these buffalo covered. The fact
that buffalo inhabited the Great Plains from Mexico to Canada is commonly known. Less well known is the fact that at the time the Pilgrims
landed at Plymouth Rock, some 2 million to 4 million buffalo lived east
of the Mississippi—in every future state but the northeast cluster of
Maine, Vermont, New Hampshire, Connecticut, and Rhode Island.[16]

The buffalo, in short, was a remarkable survivor. It had prevailed
through prehistoric episodes that forced hundreds of other mammals
into extinction. It thrived in climates ranging from subtropic Mexico to
subarctic Canada. It could defend itself against the relentless attacks of
the most deadly predators in its environment, including packs of wolves
and prehistoric humans. The buffalo, it seemed, was perfect.

By July of 1872, George Bird Grinnell and three companions
were crossing into northern Kansas on horseback, hurrying to catch up
with a tribe of 4,000 Pawnee Indians. Relations between whites and
Pawnees were peaceful, in part because the Pawnee ceded their traditional Kansas and Nebraska lands early. The Pawnee were enemies of the
Sioux and the Cheyenne and often served as guides to U.S. Army soldiers during the wars with those tribes.[17]

In 1872, the Pawnee were confined to a small reservation along
Nebraska's Loup River. Twice a year, though, the army allowed the tribe

to travel south to hunt buffalo in their historic Kansas hunting grounds. "[F]or a little while," wrote Grinnell, "they returned to the old free life of earlier years, when the land had been all their own, and they had wandered at will over the broad expanse of the rolling prairie."[18] The tribe left the reservation two weeks before Grinnell arrived in Nebraska, so now Grinnell and his companions hurried to catch up. Their guide was Lute North, who, like his brother, was an experienced young army officer. Lute spoke fluent Pawnee and was known and respected by the tribe.

After a few days they overtook the Pawnee, cresting a butte to find 200 lodges, an entire village, spread across the prairie. The head chief, Peta-la-shar, received them warmly, referring to Lute as "my son." Peta-la-shar reported that the hunt, so far, had not been successful. "But tomorrow," he promised, "a grand surround will be made." His young scouts had reported a large herd about twenty miles to the south.[19]

THE OLDEST CONFIRMED WEAPON FOUND IN NORTH AMERICA—A STONE spearhead—was discovered by archaeologists near Folsom, New Mexico, in 1927. Stone cannot be carbon-dated, but the material with which the famous "Folsom point" was discovered could be. The point was imbedded in bone—a 10,000-year-old bone of *Bison antiquus*, an extinct ancestor of the modern buffalo.[20] Humans have been hunting buffalo for a long time. Indeed, in twenty-seven of thirty-five key North American archaeological sites revealing the story of early North American human activity, buffalo bones outnumber those of any other animal.[21]

By 9,000 years ago, Folsom Man was using many of the same strategies and techniques that American Indians would later use when hunting buffalo on foot. The foundation of these strategies was an intimate understanding of buffalo behavior. Humans could not outrun the buffalo, but they could turn the defenses of the herd against itself.[22]

Many hunting strategies involved sophisticated team efforts to move the herd to a killing zone. This did not, as a general matter, mean frightening the herd into a full-fledged stampede—at least not at first. Hunters, for example, might use the smell of distant smoke to steer the herd. Sometimes a skillful hunter draped in a buffalo skin could decoy a herd, actually leading it in the chosen direction. Once moving, a herd could sometimes be steered into giant V's formed by stacked stones, mounds of dung, or, in winter, by a line marked in the snow.[23]

Ancient humans, and later Indians on foot, used different types of killing zones. The most famous is the buffalo jump, or *pishkun*. Meriwether Lewis described the critical moment: "The disguised Indian or decoy has taken care to place himself sufficiently nigh the buffaloe to be noticed by them when they take to flight and running before them they follow him in full speede to the precipice ... the (Indian) decoy in the mean time has taken care to secure himself in some cranney or crevice of the clift."[24] The buffalo, pushed by their own frantic mass, tumbled to their deaths. At one Colorado site, 193 buffalo were killed in a single hunt that took place 8,500 years ago. In Montana alone, there are more than 300 *pishkun* sites. Along a one-mile stretch at one of them, Ulm Pishkun, the buffalo bones are thirteen feet deep.

Hunters on foot also killed buffalo by driving them into box canyons or giant corrals. In the winter, hunters wearing snowshoes ran down buffalo as they floundered in deep snow. Other herds were steered onto ice, where their hooved feet could find no traction.

At some point, probably around 1,500 years ago, Plains Indians began to use the bow and arrow. This development increased the range at which buffalo could be killed but not the basic hunting techniques. As a general matter, at the time Europeans arrived in North America, Plains Indians were hunting the buffalo in ways that differed little from earlier humans, thousands of years before them. In the seventeenth century, a French fur trader named Pierre Esprit Radisson was impressed by the ability of the Sioux Indians—then inhabitants of Minnesota—to hunt the buffalo on foot.[25]

The horse changed everything.

The prototypical image of the Plains Indian is the mounted hunter-warrior, and indeed, Plains Indians were among the greatest horsemen in history. But the history of Plains Indians on horseback is surprisingly short. It was not until the 1600s that Indians of the Southern Plains first acquired horses—from Spanish remudas brought up through Mexico. And it was not until the 1700s that horses were in widespread use by Indians of the Northern Plains.[26]

The horse, in a literal way, expanded the horizon of the Native Americans. A tribe on foot might travel five or six miles in a day. A mounted tribe could easily cover twenty. Increased mobility made it easier to follow the buffalo herd and gave Indians more choice about where to live. When the tribe needed to move, the horse's ability to carry more weight also meant that more food could be preserved and transported, which in turn decreased the risk of starvation in the winter.

There were downsides to this increased mobility. More traveling meant more trespassing, and Indians fought other Indians more frequently as they defended their traditional territory. The advent of the horse also increased the dependence of Indians on the buffalo. With the ability to follow the herd and the vast food supply it represented, there was little incentive to farm—or even to hunt other types of game. Some tribes, such as the Cheyenne and the Crow, abandoned farming when they came into possession of the horse. Increased dependence on the buffalo, of course, increased the risk of starvation if the buffalo should become scarce.[27]

PAWNEE CHIEF PETA-LA-SHAR MADE GOOD ON HIS PROMISE TO GRINNELL. The day after arriving at the Pawnee camp, the young stockbroker would hunt buffalo in a classic surround. Just as significant, he would witness the rituals of the traditional Pawnee hunt.

The entire 4,000-member tribe embarked in an early morning mist, having broken camp and loaded their horses in a matter of minutes. Grinnell described the grand procession, led by "eight men, each carrying a long pole wrapped round with red and blue cloth and fantastically ornamented with feathers, which fluttered in the breeze as they were borne along." These were "buffalo sticks," treated with reverence by the tribe because "the success of the hunt was supposed to depend largely upon the respect shown to them." Behind the buffalo sticks rode thirty or forty of the tribe's most important men, "mounted on superb ponies." Grinnell was given the honor of riding in this lead group, much of the time next to Chief Peta-la-shar. Finally came the great mass of women, children, and men of lower station.[28]

Grinnell was surprised to see many men on foot, sometimes leading multiple ponies. Lute North explained that they were saving their horses "so that they might be fresh when they needed them to run buffalo."

The Pawnee had stopped at a new campsite when Grinnell noticed "a sudden bustle among the Indians." On a horizon marked by distant bluffs, a horseman appeared, riding hard toward the camp. When he arrived, the rider reported quickly to the chiefs. A large herd had been spotted, only ten miles away.

Wild excitement now filled the camp. Women began immediately to break down teepees for transportation closer to the kill site. Men, meanwhile, stripped themselves and their ponies of all superfluous weight. Grinnell quickly prepared his own horse and weapons, mounting up to regard the stunning human vista of which he was privileged to be a part:

> The scene that we now beheld was such as might have been witnessed here a hundred years ago. It is one that can never be seen again. Here were eight hundred warriors, stark naked, and mounted on naked animals. A strip of rawhide, or a lariat, knotted about the lower jaw, was all their horses' furniture. Among all these men there was not a gun nor a pistol, nor any indication that they had ever met with the white men . . . Their bows and arrows they held in their hands. Armed with these ancestral weapons, they had become once more the simple children of the plains, about to slay the wild cattle that Ti-ra´ wa had given them for food. Here was barbarism pure and simple. Here was nature.[29]

Grinnell and 800 hunters now thundered across the Kansas plains. Some of the Pawnee rode one horse while leading another, saving their best mount for the chase. The less prosperous rode double, pulling two mounts along behind. Grinnell marveled at the skill of the bareback riders, so perfectly attuned to their horses, he remembered, that the plains appeared to be "peopled with Centaurs."

Despite the excitement of the hunters, tight discipline governed their advance. At regular intervals in the front of the procession rode the "Pawnee Police," whose authority during the hunt was absolute. They set the pace, ensuring that no one dashed ahead and scared the herd. A hunter who disregarded their command might be "knocked off his horse with a club and beaten into submission without receiving any sympathy even from his best friends." Much was at stake. The food supply of the tribe for the next six months would be determined in the moments about to unfold.

Ten miles from camp, the lead riders, Grinnell among them, carefully crested a high bluff. "I see on the prairie four or five miles away clusters of dark spots that I know must be the buffalo." Close now, the hunters change course, using the line of bluffs to conceal their advance.

Finally only a single ridgeline separated the mass of hunters from the mass of their prey. "The place," remembered Grinnell, "could not have been more favorable for a surround had it been chosen for the purpose." The terrain before them consisted of an open plain, two miles wide, surrounded by high bluffs. "At least a thousand buffalo were lying down in the midst of this amphitheater."[30]

In a classic surround, Indians encircled the herd before the great charge. In this hunt, though, they would employ a variant of the strategy. All 800 hunters would ride into the herd from the same side. The objective was for the fastest riders to pass all the way through the herd, then turn back to face it. If successful, the herd too would turn—into the charging bulk of the hunters.

Behind the ridgeline, the hunters assembled in a long, crescent-shaped formation. Then over the hill they rode. "[W]hen we are within half a mile of the ruminating herd a few of them rise to their feet, and soon all spring up and stare at us for a few seconds; then down go their heads and in a dense mass they rush off toward the bluffs." The leader of the Pawnee Police gave a cry, "*Loó-ah!*"

"Like an arrow from a bow each horse darted forward," remembered Grinnell. "Now all restraint was removed, and each man might do his best." Grinnell, who had only one horse, soon fell behind the Indians on fresh mounts. Great clouds of dust quickly filled the air, along with flying pebbles and clods kicked up by fleeing hooves. As he galloped forward, Grinnell could just make out the fastest riders, disappearing into the herd. Soon he could no longer see the ground, relying completely on his horse to navigate the field, aware that falling could mean death. Halfway across the valley, Grinnell realized that some buffalo were now coming back—directly at him. The herd had been turned.[31]

"I soon found myself in the midst of a throng of buffalo, horses and Indians." Grinnell began shooting, "and to some purpose."[32] Two-

A George Catlin image of a traditional Indian buffalo hunt.
Courtesy of the Amom Carter Museum.

thousand-pound animals tumbled and skidded to the earth around him. Shooting from a galloping horse required a skilled mount, steady hands, and even steadier nerve. Riders attempted to come alongside a coursing buffalo, then aimed behind the shoulder. It was difficult and dangerous. Overzealous gunners sometimes shot each other. George Armstrong Custer, likely hunting buffalo with a pistol, once misfired during a chase and blew out the brains of his horse.

Indians usually hunted buffalo with bow and arrow. The bow was far superior in a mounted chase to a single-shot muzzle-loader, and as for repeating rifles, few Indians owned them. A skilled Indian archer could not only place his shot accurately but could do so with remarkable force: Arrows sometimes protruded out the opposite side of the buffalo and occasionally passed all the way through. A Sioux warrior named Two-Lance was once observed to shoot an arrow completely through the enormous body of a running bull. Grinnell watched the Pawnee fire "arrow after arrow in quick succession, ere long bring down the huge beasts and then turn and ride off after another."[33]

Grinnell noted how the well-trained Pawnee horses could bring their riders alongside a buffalo with no guidance whatsoever, "yet watch constantly for any indication of an intention to charge and wheel off."[34] On the return trip of the Lewis and Clark expedition, Sergeant Nathaniel Hale Pryor found it difficult to herd well-trained Indian ponies, because at every sighting of buffalo the ponies would dash off in pursuit, "surround[ing] the buffalo herd with almost as much skill as their riders could have done." In frustration, Sergeant Pryor finally resorted to sending one rider ahead of the main party to drive away all buffalo before the arrival of their horses.[35]

Grinnell's own mount, as he learned the hard way, was untrained in the hunting of buffalo. One of Grinnell's shots was off its mark, striking a cow without bringing her down. The cow spun around to make a "quick and savage charge." Grinnell's pony reacted slowly, and "his deliberation in the matter of dodging caused me an anxious second or two." The horse just missed being gored by a swipe of the cow's horn.[36]

Having filled his quota of adventure for the day, Grinnell retired to a high knoll from which he could watch the rest of the hunt and its aftermath. The plain was dotted with downed animals, each soon attended by two or three men. The women would eventually catch up to the hunt and take over the grunt work of processing the hides and the meat.

PLAINS INDIANS WERE BORN ON A BUFFALO ROBE AND WRAPPED IN A buffalo robe when they died. In between, the buffalo was the foundation of both their economy and their culture. Before the arrival of whites, buffalo provided for virtually every need.[37]

The Indians' use of nearly every part of the buffalo they killed is well known. Certain nutrient-rich organs were cut from the still-warm animals and consumed raw, including the liver and the kidney. For days after a successful hunt, the entire tribe gorged on fresh meat at celebratory feasts. Delicacies included hump meat, tongue, nose, hot marrow from the roasted bones, calf brain cooked in the skull, soup made from blood, and *boudins*—intestines filled with diced tenderloin and then boiled. During the frenzied post-hunt feasting, men might eat ten or even fifteen pounds of meat. Meat not consumed in the immediate aftermath of the hunt was dried or smoked into jerky, some of which was pounded into pemmican, a nutritious mixture of meat, fat, and berries. Both jerked meat and pemmican could be stored for months. So too the buffalo's thick back fat, which was stripped off and smoked.

Having provided the food, the buffalo also provided the fuel for cooking. As generations of white hunters and settlers would also learn, fires on the treeless plains were built with buffalo chips (dried dung).

Before the arrival on the plains of canvass as a popular trade item, Indian teepees were sewn from the hides of buffalo cows. It took around twenty for a single teepee. Cows' hides were thinner and therefore both lighter and easier to work. Hides could also be used to make clothing (though the even softer hides of deer were also popular). The thick hides of bulls were used to make rawhide, battle shields, even "bullboats"— rawhide stretched over willow branches in the shape of a giant bowl. Rawhide pulled over a wooden frame was also used to make saddles, sometimes padded with buffalo hair.

Dozens of other objects came from the buffalo. Spine sinews became thread pulled by needles made from sharp fragments of bone. Tendons were used to make bowstrings. A dried tongue worked as a comb. Hair was braided into rope. The paunch held water, even during cooking. Bones became mallets, digging tools, awls. Horns became waterproof powder horns or spoons. And the Indians used the tail, as its previous owner did, to swat flies.

THAT NIGHT AT THE CAMP, GRINNELL AND HIS COMPANIONS JOINED the Pawnees' feasting and celebration of the successful hunt. For his part, Grinnell could now say that he, like his wilderness heroes, had felled the mighty bison. Yet his mood was pensive. "And so the evening wears away, passed by our little party in the curious contemplation of a phase of life that is becoming more and more rare as the years roll by."[38]

The following year, the Pawnee returned to Kansas for their semi-annual hunt. They were attacked by the Sioux. One hundred and fifty-six Pawnee were killed, including a large number of women and children. The Pawnee would never hunt buffalo again.[39]

In the same year, 1873, Grinnell published an article about his buffalo hunt in a sporting journal called *Forest and Stream*. The article included a grim warning about the buffalo: "[T]heir days are numbered, and unless some action on this subject is speedily taken not only by the States and Territories, but by the National Government, these shaggy brown beasts, these cattle upon a thousand hills, will ere long be among the things of the past."[40]

It was a remarkable flash of foresight, written at a time when, by Grinnell's own description, buffalo still "blackened the plains." Yet with all his prescience, even Grinnell could little imagine the power of the forces then conspiring against the buffalo. Change, sudden and dramatic, was on the very near horizon.

CHAPTER FOUR

"I Felled a Mighty Bison"

This evening, about 5 o'clock, I felled a mighty bison to the earth. I placed my foot upon his neck of strength and looked around, but in vain, for some witness of my first great "coup." I thought myself larger than a dozen men.

—WILLIAM MARSHALL ANDERSON,
1834[1]

In 1519, Hernando Cortez and a handful of Spanish legionnaires entered the great Aztec city of Tenochtitlan (now Mexico City), capital of the Aztec Empire as well as home of its emperor—Montezuma II. Montezuma's city was an alien wonderland, including a menagerie of exotic animals given to the emperor as gifts or captured by his hunters. According to Spanish historian Antonio de Solis y Ribadeneyra, Cortez and his men saw "Lions, Tygers, Bears, and all others of the savage kind which New-Spain produced." Of all Montezuma's beasts, however, "the greatest Rarity was the Mexican Bull, a wonderful composition of divers Animals." The animal "has crooked shoulders, with a Bunch on its back like a Camel; its Flanks dry, its Tail Large, and its Neck covered with Hair like a Lion. It is cloven footed, its Head armed like that of a Bull, which it resembles in Fierceness, with no less Strength and Agility."[2]

The first Europeans to see the buffalo in its native habitat were part of another small band of Spaniards, including a man named Alvar Nuñez Cabeza de Vaca. Shipwrecked in 1530 on the Gulf Coast of present-day Texas, Cabeza de Vaca encountered what he called cattle. In a book about his adventures, he described in detail an Indian tribe he dubbed the "Cow nation," who hunted the buffalo and then distributed "a vast many hides into the interior country." As a food, Cabeza de Vaca found buffalo to be "finer and fatter" than the beef of his native Spain.[3]

The first Englishman to see the buffalo was an explorer named Samuel Argall, who in 1612 sailed a small frigate to the navigable headwaters of the Potomac River. There, with a group of his crewmen, he went "marching into the Countrie," where he found a "great store of cattle as big as Kine [oxen], of which the Indians that were my guides killed a couple, which we found to be very good and wholesome meate, and very easy to be killed, in regard they are heavy, slow, and not so wild as other beasts of the wildernesse." Assuming that Argall and his men did not penetrate far into the thick woodlands that hemmed the Potomac, it is quite likely that the buffalo he described were within the boundaries of what is today Washington, D.C.[4]

Though less associated with the frontier territory east of the Mississippi, buffalo appear in numerous historical accounts. In 1701, a colony of Huguenots on the James River attempted, unsuccessfully, to domesticate two captured calves. In a 1733 report, George Oglethorpe, the first governor of Georgia, listed buffalo among the wild animals of his colony. Early settlers to hunt buffalo included residents of Pennsylvania, West Virginia, Ohio, Indiana, and Illinois. When Daniel Boone and other trailblazers penetrated the continent westward into Kentucky and Tennessee, they followed paths first trodden by buffalo, heading east.

Though the story of the buffalo in the eastern United States is less chronicled, the outcome, in its essence, rings familiar: By 1800, the buffalo east of the Mississippi had been exterminated.[5]

IN THE LOUISIANA PURCHASE OF 1803, NAPOLEON BONAPARTE SOLD THE United States a fresh herd of buffalo, along with the half-billion pristine acres on which they roamed. Meriwether Lewis and William Clark became the first Americans to explore the newly acquired land. Generations of American schoolchildren have committed to memory the goal of their exploration of the American West: to discover a passage by water to the Pacific Ocean. But President Thomas Jefferson gave more specific directions in his 1803 written orders to Captain Lewis:

The object of your mission is to explore the Missouri river, &
such principal stream of it, as, by it's course and communication
with the waters of the Pacific ocean, whether the Columbia,
Oregan, Colorado or any other river may offer the most direct
& practicable water communication across this continent for the
purposes of commerce.[6]

As the men of the Lewis and Clark expedition crossed the great
western desert, the buffalo played a vital role in their very survival. Lewis
and Clark's combined journals discuss the buffalo in at least 707 separate
entries. The first mention occurred on June 6, 1804, in present-day Missouri, when Clark noted that they observed "Some buffalow Sign today."
On August 23, 1804, the expedition killed its first bison near the
Nebraska–South Dakota border.[7]

As with later travelers in the West, the buffalo provided the Lewis and
Clark expedition with a vital source of food. When it was available—
most particularly as they crossed Montana between Fort Mandan and
the Continental Divide—the men ate buffalo, and in prodigious quantities. "The hunters killed three buffaloe today," wrote Captain Lewis on
July 13, 1805, near the Great Falls of the Missouri River. "[W]e eat an
emensity of meat; it requires 4 deer, an Elk and a deer, or one buffaloe,
to supply us plentifully 24 hours." A typical buffalo cow—always preferred over bulls when it came to eating—provided around 400 pounds
of meat. Given the opportunity, the thirty-seven men of the expedition
could consume 9 pounds of fresh meat per day per man.

If the voyage of Lewis and Clark famously failed in its mission of
finding a water passage to the Pacific, it certainly succeeded in opening up new streams of commerce. Commerce, in the context of the far
West of the early nineteenth century, meant fur. Fur, as a general matter, meant beaver. Hence the importance of Captain Lewis's famous
report back to President Jefferson at the conclusion of the voyage in
1806: "The Missouri and all it's branches from the Chyenne upwards
abound more in beaver and Common Otter, than any other streams
on earth, particularly that proportion of them lying within the Rocky
Mountains."[8]

The period of Western history between 1806 (Lewis and Clark's return) and 1841 (the start of the great California–Oregon migration) was defined by the pursuit of beaver.

Since the mid-1600s, fashionable Europeans had worn hats they called beavers. The beaver hat was made not from the animal's skin (à la the coonskin cap) but rather from the downy underlayer of shorthairs. These shorthairs were trimmed from the skin, chemically treated, and pressed into felt. The felt was used to make the hat. By 1800, the European beaver had been trapped into near extinction, placing a particular premium on the North American trade.

So central was the beaver to the fur trade that the Hudson's Bay Company, operating continuously in Canada since 1670, used beaver pelts as the currency of exchange. "Beaver being the Chief Commodity we Trade for," wrote a company officer, "We therefore make it the Standard whereby we value all Furs and Commodities." In 1811, for example, it took almost three buffalo robes to equal the value of a single beaver pelt.[9]

Beaver pelts, in sharp contrast with buffalo hides, were well matched to both the consumer demands and the trade logistics of the day. While the demand for beaver felt hats was high, the demand for buffalo hides was limited. The leather made from buffalo hides was notoriously soft and spongy, unusable for most common applications (such as the making of shoes and belts). The primary application for buffalo was robes used as cold-weather covering by riders in sleighs and carriages. Another common use was among teamsters: The famous warmth of buffalo coats made them a veritable symbol of the trade, though their great weight made them impracticable for the average person on foot.[10] Aside from these limited areas, demand for buffalo robes was modest.

From the standpoint of logistics, beaver pelts had another great advantage over buffalo hides. A beaver pelt was lightweight, barely more than a pound, and therefore relatively easy to transport. The hide of a bull buffalo, when first stripped from the carcass, could weigh as much as 150 pounds. The process of staking out and drying the hide caused it to lose around two-thirds of its weight, but even a "flint" hide could tip the scales at 50 pounds. The difficulty of transporting such heavy goods ate quickly into profits—even for those locations with access to the Missouri River, the principal highway for robes before the advent of the railroad.

By comparison with what was to come, the trade in buffalo hides prior to the 1870s was modest. This was particularly true in the years before 1840, when beaver provided an attractive alternative. In this era, most buffalo robes were supplied not by white hunters but in trade with the Indians. Coffee, sugar, calico, blankets, butcher knives, beads, guns, and whiskey were common items of exchange.[11]

By 1840, the beaver trade had come crashing to the ground. On the supply side, the beaver had been plucked clean from virtually every waterway that drained the Rocky Mountains. On the demand side, meanwhile, fashion proved fickle. Silk top hats became the rage, and the market for beaver pelts evaporated (just in time, in all likelihood, to save the humble animal from extinction).

With the beaver trade gone, some former trappers turned to the hunting of buffalo. Still, neither the lack of demand for buffalo nor the difficulties in transport had been solved in the period between 1840 and 1870. The available numbers are haphazard. But even with the new interest in buffalo resulting from the end of the beaver era, it appears that only rarely would the annual harvest exceed 200,000 hides. Many years produced fewer than 100,000. In the late 1830s and early 1840s, there is a revealing series of letters between Pierre Chouteau, a leading trader in St. Louis, and Ramsay Crooks, president of the American Fur Company (the largest producer of buffalo hides). In their correspondence, the two men worry constantly about flooding the buffalo hide market even at modest levels of trade, and Crooks repeatedly directs Chouteau to send fewer robes east.[12]

It was about this time, 1843, that the painter John James Audubon made his last journey west. Beginning from his home in Audubon Park, Audubon ultimately ascended the Missouri and then the Yellowstone River through present-day Montana. At a time when few prairie travelers could see past a landscape covered in buffalo, Audubon (like his wife's student, George Bird Grinnell), had the rare ability to see over the horizon. "One can hardly conceive how it happens," wrote Audubon in his diary, "so many [buffalo] are yet to be found. Daily we see so many that we hardly notice them more than the cattle in our pastures about our homes. But this cannot last; even now there is a perceptible difference in the size of the herds, and before many years the Buffalo, like the

Great Auk, will have disappeared."[13] (Grinnell would read Audubon's journal before his first trip west.)

It was a flash of remarkable insight, for even as Audubon spoke, the pace of change had begun to accelerate.

IN 1836, A PARTY OF FIVE MISSIONARIES CROSSED THE CONTINENT with the intention of ministering to the native tribes of the Pacific Northwest. The party traveled in the company of a large group of fur trappers and included the first two white women ever to cross the Rockies—Narcissa Whitman and Eliza Spalding. Eighteen-year-old Narcissa kept a diary, and the experiences she recorded foreshadowed a generation of emigrants on the verge of transforming the American West. The buffalo played a central role in their drama.

For emigrants traveling west, sighting the first buffalo marked a signature moment in their voyage—true arrival on the frontier. For Narcissa, the event took place on June 4, 1836, in the Platte Valley of Nebraska, thirty miles above the confluence of the river's north and south branches.

> We have seen wonders this forenoon. Herds of buffalo hove in
> sight; one a bull, crossed our trail and ran upon the bluffs near
> the rear of the camp. We took the trouble to chase him so as
> to have a near view. Sister Spalding and myself got out of the
> wagon and ran upon the bluff to see him. This band was quite
> willing to gratify our curiosity, seeing it was the first. Several
> have been killed this forenoon. The [fur trade] Company keep[s]
> a man out all the time to hunt for the camp.[14]

As with Lewis and Clark, the buffalo provided early transcontinental emigrants with a vital source of food. Indeed provisions for the voyage were planned around the expectation of reaching the herd. "On the way to buffalo country we had to bake bread for ten persons," wrote Narcissa. "It was difficult at first, as we did not understand working outdoors . . . June found us ready to receive our first taste of buffalo."[15]

And she liked it. Reporting rapturously on their new food source, Narcissa wrote, "I never saw anything like buffalo meat to satisfy hunger.

We do not want anything else with it. I have eaten three meals of it and it relishes well." Good thing she liked it: "We have had no bread since [reaching buffalo country]. We have meat and tea in the morn, and tea and meat at noon. . . . So long as there is buffalo meat I do not wish anything else."[16]

Like other travelers on the treeless plains, Narcissa discovered that the buffalo supplied not only the meat but also the means to cook it. "Our fuel for cooking since we left timber (no timber except on rivers) has been dried buffalo dung; we now find plenty of it and it answers a very good purpose, similar to the kind of coal used in Pennsylvania." Anticipating the reaction of her eastern relatives to the notion of cooking with dung, Narcissa noted, "I suppose now Harriet will make up a face at this, but if she was here she would be glad to have her supper cooked at any rate in this scarce timber company." Some pioneers, putting the best possible sheen on their fuel source, dubbed it *bois de vache*—wood of the cow.[17]

After six weeks of a diet consisting exclusively of buffalo roasted over buffalo dung, Narcissa's enthusiasm finally began to fade. "I thought of mother's bread, as a child would, but did not find it on the table," she wrote on July 18. "I should relish it extremely well; have been living on buffalo meat until I am cloyed with it."

Still, as Narcissa would soon discover, fresh buffalo meat, however tiresome, was preferable to the available alternatives. As they traveled farther west, buffalo became scarce. "Have seen no buffalo since we left the [Green River] Rendezvous," she wrote on July 27. "We have plenty of dried buffalo meat, which we have purchased from the Indians—and dry it is for me. It appears so filthy! I can scarcely eat it; but it keeps us alive, and we ought to be thankful for it."

In late July or early August, Narcissa and her fellow travelers reached Fort Hall in Idaho. At the fort they ate more dried buffalo, along with "mountain bread," described in Narcissa's journal as "course flour and water mixed and roasted or fried in buffalo grease." Not her mother's bread, but "[t]o one who has had nothing but meat for a long time, this relishes well."[18] This is the last mention of buffalo in her journal.

A few weeks later, Narcissa and her husband, Marcus Whitman, settled in the Walla Walla River Valley, founding a mission and ministering to the Cayuse Indians. After initial enthusiasm, events began to degenerate. The Whitmans lost their two-year-old daughter in a drowning accident. Relations with the Indians spiraled downward, especially after a measles epidemic killed most of the children in the Cayuse tribe. In 1847, a band of Cayuse murdered Narcissa, Marcus, and a dozen other whites, then burned their mission to the ground.[19]

If the Whitmans failed in conveying their religious message to the Indians, they succeeded in conveying a message to their fellow whites, east of the Mississippi: White women—white *families*—could cross the continent. Indeed a wagon could be pulled across most of it. Unwittingly, the Whitmans were at the vanguard of America's great westward migration.

In 1841, sixty-nine emigrants—including five women—crossed the continent by wagon. Their original goal was California, though half of the party ultimately would decide to veer north for Oregon. In 1842, a group of about 100 ventured to Oregon. The year 1843 marked the first large-scale crossing: 1,000 emigrants herding 5,000 head of cattle.[20]

As it had for Narcissa Whitman, the buffalo played a central role in the frontier experience of other early emigrants—both symbolically and practically. If spotting the first buffalo signified arrival on the frontier, killing a buffalo—for food or otherwise—was a veritable right of passage for any man who traveled west. William Marshall Anderson, traveling near the present-day Nebraska–Wyoming border, recorded his first kill with near biblical zeal. "This evening, about 5 o'clock, I felled a mighty bison to the earth. I placed my foot upon his neck of strength and looked around, but in vain, for some witness of my first great 'coup.' I thought myself larger than a dozen men."[21]

A powerful psychology took grip that no man seemed able to resist. Captain Howard Stansbury, who led a group of emigrants near the South Platte, recorded the following after the party's first encounter with a large herd of buffalo: "The effect upon our hunters, and, in fact, upon the whole party, was that of a sudden and most intense excitement, and a yearning, feverish desire to secure as much as possible of this noble game." A pioneer named Josiah Gregg described the scene after his party sighted their first herd:

*Pell-mell we charged the huge monsters, and poured in a brisk
fire, which sounded like an opening battle; our horses were
wild with excitement and fright; the balls flew at random . . .
One [buffalo] was brought to bay by whole volleys of shots; his
eyeballs glared; he bore his tufted tail aloft like a black flag; then
shaking his vast and shaggy main in impotent defiance, he sank
majestically to the earth, under twenty bleeding wounds.*

In summarizing the iron grip of this buffalo fever, Gregg stated simply that "[s]uch is the excitement that generally prevails at the sight of
these fat denizens of the prairies that very few hunters appear able to
refrain from shooting as long as the game remains within reach of their
rifle." Such hunters were called "plinkers" for the potshots they fired as
they rolled along the trail.[22]

Once the initial excitement faded, some pioneers found the buffalo
so numerous as to be an irritant. Women waved aprons to shoo the
beasts away from cooking fires, and wagon trains sometimes were delayed
for days at a time by herds that refused to be hazed from the path.[23] A
pioneer named Obadiah Oakley, who traveled to Oregon from Peoria,
described herds "as thick as sheep ever seen in a field" and complained
that animals "moved not until the caravan was within ten feet of them.
They would then rise and flee at random, greatly affrighted, and snort-
ing and bellowing to the equal alarm of the horses and mules."[24]

When not alarmed, horses and mules sometimes fell under the spell
of their wild brethren and ran off with the buffalo. Explorer John Fre-
mont described how one of his mules "took a freak into his head, and
joined a neighboring band of buffalo today. As we were not in condition to lose horses, I sent several of the men in pursuit . . . but we did not
see him again."[25]

In the early years of western emigration, the numbers, at least, still
favored the buffalo.

A total of no more than 10,000 pioneers traveled to Oregon and
California during the years 1841 to 1846.[26]

Soon though, successive waves of enticements began luring more
and more Americans west. In 1847, Mormon leader Brigham Young

A pioneer woman with a load of *bois de vache*.
Courtesy of the Kansas State Historical Society.

established Salt Lake City as a western haven for his beleaguered tribe. In 1848 came the earth-shaking discovery of gold at Sutter's Mill in California. In the early 1850s, Congress began to give away free land to encourage settlement in the territories of Oregon and Washington. This modest homesteading program was expanded dramatically when President Lincoln, on May 20, 1862, fulfilled a campaign promise by signing the Homestead Act. Six weeks later, even as his army failed in an attempt to end the Civil War by taking the Confederate capital of Richmond, Lincoln affixed his signature to another bill. The law it created would transform the West to a greater degree than any measure that came before it. From two poles, Sacramento and Omaha, began the construction of the Union Pacific Railroad.

IN 1869, LESS THAN SEVEN YEARS AFTER LINCOLN LAUNCHED THE building of the transcontinental railroad, trains from east and west came together on a windswept butte in Utah. Leland Stanford, an executive of the Central Pacific Railroad, drove a golden spike into a waiting tie, and with this final fall of the sledge, America's manifest destiny had been fulfilled. East and West had been connected.

In addition to the main line between Omaha and Sacramento, numerous other lines were being built or planned. In 1870, the Kansas Pacific Railroad connected Kansas City and Denver. In the same year, construction began on the Northern Pacific, a railroad to connect the Great Lakes with the Puget Sound. All along the main Union Pacific, meanwhile, rails branched out in the form of spur lines to connect the dots bypassed in the first wave of construction. All told, a remarkable 35,000 miles of new track were laid between 1866 and 1873.[27]

Certainly the benefits of the new railroads were incalculable. Psychologically, as historian Stephen Ambrose pointed out, the Civil War had united the country North and South, but it took the completion of the railroad to bind the nation East and West. A more quotidian benefit, of course, was the dramatic decrease in the time and expense of transportation. Before the railroad, coast-to-coast transportation by wagon was measured in months of arduous travel. By train, the journey took seven days. In the wagon train era, pioneers might pay $1,000 for the equipment and provisions to cross the continent. In June of 1870, a third-class "emigrant" ticket cost $65. For any location reachable by train, wagon travel virtually ceased. As George Bird Grinnell discovered in his trip west only fourteen months after the driving of the golden spike, the old pioneer trail already had grown over with flowers for lack of traffic.[28]

Few changes so dramatic, however, come without a price. Arthur Ferguson, a young man who thought about the future as he worked to survey the route for the railroad, wrote in his 1868 journal that "[t]he time is coming and fast too, when in the sense it is now understood, THERE WILL BE NO WEST."[29] A bit of hyperbole, perhaps. Yet clearly the railroad accelerated the transformations—positive and negative—of the nineteenth century.

FOR THE GREAT HERD OF BUFFALO THAT ONCE ROAMED THE PLAINS IN AN unbroken mass from Mexico to Canada, the impact of humankind had been significant even before the earliest waves of California and Oregon emigrants. John James Audubon was not the only early western traveler to notice the diminishing numbers of the herd. A trapper named Osborne Russell kept a journal from 1834 to 1843. Writing about the buffalo, he warned that "it will not be doubted for a moment that this noble race of animals, so useful in supplying the wants of man, will at no far distant period become extinct in North America."[30] Painter George Catlin, whose dramatic images helped to create the nation's visual consciousness of the West, also warned of the buffalo's demise in the 1830s. "It is truly a melancholy contemplation for the traveler in this country, to anticipate the period which is not far distant, when the last of these noble animals, at the hands of white and red men, will fall victims to their cruel and improvident rapacity." [31]

Catlin's journal—consistent with other contemporaneous documents—hints at one impact on the buffalo herd that fits uncomfortably with our modern, popular images. Native Americans made their own contribution to the demise of the buffalo. While the image of Indians using every bit of the buffalo they killed appears far more common than wanton slaughter, wasteful killing by Indians did take place. Catlin's journal, for example, recounts an 1832 incident in which 500 Sioux killed 1,400 buffalo solely for their tongues—to be traded for whiskey with the American Fur Company.[32]

More significant than such isolated examples of buffalo slaughter by Native Americans was the impact of technology—the horse—on the Indians' hunting techniques. Hunting on horseback allowed far greater choice in target selection than hunting on foot, and given the choice, hunters shot cows over bulls. The cows' meat was superior and their hides easier to work. By the 1860s, anecdotal reports indicated cow-to-bull ratios of as high as 10 to 1. Modern game laws, of course, seek precisely the opposite effect—protecting cows instead of bulls. The impact of selective hunting, according to biologist Dale Lott, "sent the population in a downward spiral."[33]

The westward emigration that began in the early 1840s caused the buffalo to retreat from the travel corridor of the California–Oregon Trail. While early pioneers could depend on buffalo meat as a food source when they crossed the Nebraska and Wyoming plains, emigrants by the 1850s were forced to rely on bacon. The combination of plinking and subsistence hunting drove the buffalo away from the trail, but there was another factor as well. The emigrants' stock competed with the buffalo herd for grass. By the end of the wagon train era, pioneers were sometimes forced to drive their animals as far as eight miles off the trail to find suitable forage. Such denuded land offered little attraction for the herd.[34]

Construction of the railroad sealed the fate of the buffalo. If the impact of emigrants was significant, the disruption caused by the army of men who built the railroad was even greater. The construction crews that built the Kansas Pacific, for example, numbered about 1,200 men. To feed these workers, the railroad hired a young hunter named William Cody. In the company of a horse he named Brigham and a gun he named Lucretia Borgia, Cody killed 4,280 buffalo in eighteen months of Kansas service.[35]

It was in this same era that Cody won his nickname, Buffalo Bill. When it was discovered that an army scout, Billy Comstock, went by the same moniker, a contest was demanded to settle the title. "We were to hunt one day of eight hours," remembered Cody in his autobiography. "The wager was five hundred dollars a side, and the man who should kill the greater number of buffaloes from on horseback was to be declared the winner." The newspapers loved it, stoking the fires of controversy. The Kansas Pacific even sent out an excursion train full of spectators from St. Louis to witness the showdown. In Kansas (unlike the North Platte Valley of Nebraska), buffalo were still commonplace along the tracks. The excursion train pulled up alongside a suitable herd some twenty miles east of Sheridan, and the spectators spilled out, carrying picnic baskets and bottles of cold champagne. After three runs through the herd (with the occasional break for champagne), Cody beat Comstock by a score of 69 to 46. The Kansas Pacific gathered up the best heads, mounted them, and put them on display in rail stations around the country.[36]

Shooting buffalo from a moving train (like its antecedent, shooting from the deck of the steamboat) was a popular sport, while it lasted. An 1869 article in *Harper's New Monthly Magazine* described a scene from a train ride between Denver and Salt Lake City: "It would seem to be hardly possible to imagine a more novel sight than a small band of buffalo loping along within a few hundred feet of a railroad train in rapid motion, while the passengers are engaged in shooting, from every available window, with rifles, carbines, and revolvers. An American scene, certainly."[37]

The Northern Pacific Railroad in Montana offered its passengers the opportunity to "test the accuracy of their six-shooters by firing at the retreating herd." The Kansas Pacific once chartered a buffalo excursion to a church group, including 26 representatives of the fairer sex. A reporter for *Frank Leslie's Illustrated Newspaper* documented the hunt, which culminated in the killing of a bull. "[A] rope was attached to his horns, and two long files of men, with joined hands, and preceded by the band, playing Yankee Doodle, dragged him bodily to the front car and hoisted him aboard."[38]

THE CONSTRUCTION OF THE TRANSCONTINENTAL RAILROAD MADE permanent a new geographic distribution of the American buffalo. Instead of a great mass stretching unbroken across the plains from Mexico to Canada, now there were two herds—northern and southern. The southern herd was larger, encompassing the present-day states of Nebraska, Kansas, Colorado, Oklahoma, and Texas. The smaller northern herd was spread across Wyoming, the Dakotas, and Montana.

In the three centuries since the arrival of the Europeans on the North American continent, the buffalo had been winnowed dramatically, though an estimate of the decrease is difficult. One scientist who studied the numbers came to this conclusion: "One may assume with reasonable certainty that the bison population west of the Mississippi River at the close of the Civil War numbered in the millions, probably in the tens of millions. Any greater accuracy seems impossible."[39]

Despite a significant decrease in population, the idea that the buffalo could become extinct still failed to find purchase among average Americans of the early 1870s. The myth of inexhaustibility, by contrast, found support in the descriptions of surviving herds that still "blackened the prairie." Indeed the ungraspable vastness of the prairie itself lent credibility to the notion that some infinite number of buffalo must surely remain beyond the reach of mankind.

But the railroad, as it spread across the nation, was making the country smaller. Of the two major impediments to wholesale harvest of the buffalo—market demand and efficient transportation—the railroad had conquered one.

CHAPTER FIVE

"The Guns of Other Hunters"

*Often while hunting these animals as a business, I fully
realized the cruelty of slaying the poor creatures. Many
times did I "swear off," and fully determine I would break
my gun over a wagon-wheel when I arrived at camp. . . .
The next morning I would hear the guns of other hunters
booming in all directions and would make up my mind
that even if I did not kill any more, the buffalo would soon
all be slain just the same . . .*

—CHARLES "BUFFALO" JONES[1]

In the fall of 1870, a 19-year-old Vermonter named J. Wright Mooar
left his home, or as he put it, "turned my face west." He traveled to
Fort Hays, Kansas, and took a job cutting wood for the army. By 1871,
Mooar was working as a hunter, shooting buffalo to supply meat for the
fort. In the winter of 1871–1872, a local fur trader spread the word of an
intriguing proposition from Europe: An English company wanted 500
buffalo hides in order to experiment in the making of leather. Mooar
signed on as one of the hunters and helped to supply the hides.

After filling his share of the English order, J. Wright found himself
with fifty-seven surplus skins. He arranged to send them to his brother
John in New York City, telling him of the English experiment and urging him to find a New England tanner with similar interest. The hides
were transported through the streets of New York City on an open
wagon, and J. Wright described how "the novelty of the sight created a

diversion that amounted to a mild sensation." Before the hides had even been delivered to John, several fur dealers were in tow. "[T]hose 57 hides were sold to the tanners, made up into leather, and the experiment proved immediately successful."

For tanners, the timing could not have been more opportune. The industrial revolution of post–Civil War America had resulted in an explosion in the manufacturing of all manner of heavy machinery, and the belts that turned the wheels of those whirring machines were made from leather. With the discovery of new processing techniques, buffalo leather could now—for the first time—be used for the same purposes as cow leather. Demand exploded.

As J. Wright Mooar remembered, "even before the English firm had reported its success in the treatment of the buffalo hides ... I was apprised of the fact that the American tanners were ready to open negotiations for all the buffalo hides I could deliver." Mooar's brother and cousin quit their New York City jobs and headed west to join him in this promising new venture.[2]

The last levy had been breached.

THE HUNTERS WHO CAME TO KANSAS IN THE EARLY 1870S, LIKE J. WRIGHT and John Mooar, were young men seeking money and adventure. Many were Civil War veterans who found civilian life too dull for their liking. A hunter named Frank Mayer, barely a teenager during the war, had served as a bugle boy in the Confederate Army of his native Louisiana. "Fortunately for us then we had what you don't have now," remembered Mayer in his memoirs. (He died in 1954 at the age of 104.) "We had a frontier to conquer. It was a very good substitute for war."[3]

Blunt and concise, Mayer explained the logic that compelled him to chase buffalo: "He had a hide. The hide was worth money. I was young, 22. I could shoot. I liked to hunt. I needed adventure. Here it was. Wouldn't you have done the same thing if you had been in my place?"[4] Certainly thousands of Mayer's contemporaries did.

Mayer was lucky enough to learn the ropes from Bob McRae and Alex Vimy, two experienced frontiersmen he met in Dodge City. McRae's frontier credentials included punching cows and dealing faro. According to Mayer, "He ran away from a nagging wife to the quietude of the buffalo ranges." Alex Vimy was a "French-Indian breed" who was the "best knife and tomahawk thrower in the whole southwest." Vimy was also on the lam, having "knifed a lumberjack in a squabble over a girl." There had been other knife fights too. In his pocket, as a good-luck charm, Vimy carried the ear of another former rival. Mayer claimed that McRae and Vimy were typical of the buffalo hunters he knew. If so, it's not hard to understand why he called them the "saltiest goddamn men on the Western frontier."[5]

Their past lives may have qualified them as "salty," but one experience few could claim, circa 1872, was professional buffalo hunting. Aside from a few grizzled veterans of the Missouri fur trade, virtually no one had ever made his living from selling hides.

So they learned their trade as they went along. Because shooting from the ground was regarded as lowbrow, some early hunters attempted to harvest buffalo from horseback. As Buffalo Bill demonstrated, the mounted chase made for great spectacle—but as a business model it was a bust. The mounted hunter scattered the herd and left dead buffalo along a trail that could stretch for miles. Gathering up the hides of these far-flung victims was inefficient. By the time a skinning wagon had been dragged from carcass to carcass, no profit remained.[6] Still, so powerful was the romantic image of running buffalo that hunters insisted on calling themselves "buffalo runners," despite the fact that they hunted on foot. "Why a runner?" asked Frank Mayer when he first arrived in Kansas. "We don't run buffalo the way you kill them." His partner's reply: "Must be because we have to do a hell of a lot of running across the plains to find them."[7]

A few early hunters experimented with variants of the old Indian-style surround, but that didn't suit their purposes, either. An effective surround required an army of men, diluting the proceeds in the process.[8]

Many early buffalo hunters settled on a system known derisively as "tail hunting." In tail hunting, a few men would creep up on the herd afoot, fire off as many shots as they could, run after the fleeing tails, and squeeze off a couple more rounds. "This system was afterwards dubbed 'tenderfoot' hunting," remembered a runner named Skelton Glenn, "and did not often pay expenses for either hunter or skinner."[9]

Frank Mayer described the frustrations of all the novice buffalo hunters when he said, "I was a businessman. And I had to learn a business-man's way of harvesting the buffalo crop."

What they lacked in success, they compensated in that great common denominator of all who sought their fortunes in the West—hope. As Frank Mayer put it, "always was that dream that next season, I'd recoup my losses, and make that fickle jade Fortune stand and deliver what I had coming to me." Mayer expected that he might kill a hundred buffalo a day at $3 per hide. After netting out expenses, he anticipated profits of $6,000 a month, or as he calculated, three times the salary of the president of the United States.[10] A few men did get rich selling hides, but Fortune never did deliver for Mayer. According to his careful records, he netted $3,124 in his best year—a substantial sum (though substantially less than the president's salary). After nine years on the buffalo plains, Mayer had $5,000 in the bank. In his experience, a typical hunter was lucky to clear $1,000 year, roughly equivalent to the wage of an underground miner.[11]

Hope, though, was a powerful magnet. It is impossible to know the precise number of hunters who swarmed the Kansas plains in 1872, but to Frank Mayer it seemed that "[t]he whole Western country went buffalo-wild." One estimate puts the figure between 10,000 and 20,000.[12] Whatever the true total, one thing is certain: It would soon grow much larger. In the East, financial calamity was about to rock the lives of many young men, among them George Bird Grinnell.

THE FRENETIC, PHYSICAL CONSTRUCTION OF THE RAILROADS IN THE decade before 1873 was supported by equally frenetic activity in America's financial markets. Railroads were the growth industry of their day, with rampant, can't-miss speculation in hundreds of companies.

The investment firm owned by George Bird Grinnell's father (Geo. B. Grinnell and Company, where 24-year-old Grinnell continued to grind out his days) was typical of its time. In other words, its clients were heavily invested in railroad stocks. Many of these clients traded on the margin, meaning that they purchased their stocks with money they borrowed from Geo. B. Grinnell and Company. For Grinnell's father, the risk of extending such loans was tempered by the vast resources of his partner, Horace F. Clark. Clark was the son-in-law of the firm's biggest client, Cornelius Vanderbilt. "So long as Mr. Clark was living," remembered young Grinnell, "any additional margin required was always forthcoming."[13]

In the spring of 1873, Clark died. In the fall of 1873, the mighty investment firm of Jay Cooke & Company, principal financial agent in the building of the Northern Pacific Railroad, declared bankruptcy. Panic—the Panic of 1873—ensued. In an economic depression that would stretch for six years, thousands of businesses failed and the unemployment rate soared to 14 percent.[14]

At the firm of Geo. B. Grinnell and Company,[15] Grinnell's father sought desperately to collect on the money he had lent to his clients. It was not forthcoming, in part because Cornelius Vanderbilt—who, as Grinnell noted, "might have helped"—bore a grudge against one of the other partners in the firm. Geo. B. Grinnell and Company failed, spending the next several months in contentious bankruptcy proceedings. In his memoirs, Grinnell would remember that the "winter was one of great suffering for the family."[16]

To his credit, Grinnell's father ultimately succeeded in recovering a portion of the funds owed him. In an action unique at the time, he also repaid his own debts in full and managed to leave the firm with "a little capital remaining." Exhausted, he retired to Audubon Park, leaving the business to his son.[17]

For young Grinnell, though, the whole affair had only underscored his "settled dislike for the business." After his father retired, "there was nothing to hold me to Wall Street." He shuttered the company and moved to New Haven, there to resume his Yale studies under Professor Othniel Marsh.[18]

Grinnell was not the only young American to seek out a new career amid the turmoil of the Panic of 1873. Thousands of young men with less means would head west to hunt buffalo. "They were walking gold pieces," said Frank Mayer, "and a young fellow who had guts and gumption could make his fortune."[19]

BY 1873, THE BUFFALO HUNTERS ON KANSAS'S BROAD PLAINS WERE refining their craft. As time passed, the haphazard gave way to the methodical. In the space of a few short months, the men of Kansas had created an organized industry: commercial hunting for the hides of buffalo.

The basic work unit of the market hunters evolved into "outfits" that typically consisted of one or two riflemen, a couple of skinners for each shooter, and one or two men to cook and keep camp.[20]

Their equipment, though simple, was expensive. As Frank Mayer explained, "over 10,000 men were bidding against one another for the necessities of the life they had chosen." Mayer spent his life savings, $2,000, on two wagons, eighteen mules, rifles, ammunition, bedrolls, tents, and cooking utensils. The high cost didn't worry him, though. "I'd make it up the first month, I kept telling myself, the very first month."[21]

All self-respecting hunters made their own ammunition. Factory-loaded cartridges were expensive—25 cents a round versus half that price for hand-loaded. The runners anticipated shooting a lot, provisioning themselves with a veritable armory's worth of supplies. John Cloud Jacobs, who chased buffalo in Texas, began the season with a ton of ammo, literally: "sixteen hundred pounds of lead, and four hundred pounds of powder—beside shells, paper caps, etc." Jacobs described how the bullets he molded were inspected for the "least bit of flaw." If imperfect, the offending ball was "put back in the heat." Frank Mayer mixed one part tin to sixteen parts lead, which he said gave his bullets "just enough hardness to penetrate and enough lead softness to mushroom."

Men who shot guns for a living could discriminate the subtleties between different varieties of gunpowder. Mayer, for example, disliked the leading American brands, DuPont and Hazard, which he believed left a residue that was particularly difficult to clean. When it was available, he paid premium prices for imported powder from England. Paying extra, in the view of most buffalo runners, was a superior option to the alternative. "It frequently happened that a man's life depended on a cartridge that neither snapped nor flickered."[22]

The expense of ammunition and the toil of producing it underscored the importance of efficient shooting. As Frank Mayer explained, "the thing we had to have, we business men with rifles, was one-shot kills." Yet the one piece of equipment that early buffalo hunters lacked was the right gun for the job.

Rifle technology had catapulted forward during the Civil War. Advances included breech loading, metal cartridges, repeaters, and telescopic sights. The fine weapons widely available in the early 1870s included such storied rifles as Henrys, Spencers, and Winchesters. But what these guns could not do—at least not with the mass-production consistency demanded by runners—was kill buffalo at long range with a single shot.

John Mooar, who along with his brother J. Wright had midwifed the industry of commercial hunting, can also be found in the lineage of the gun that would propel the industry to its zenith. In the earliest days of the Kansas hide hunt, Mooar wrote a letter to the Sharps Rifle Manufacturing Company of Bridgeport, Connecticut. Sharps rifles were highly respected, but Mooar asked for a version built specifically for the exigencies of hunting buffalo on the plains.[23]

The Sharps Company was quick to oblige, and its Sharps buffalo rifle was the answer to the prayers of the Mooar brothers and their fellow hunters. Indeed the men who carried the Sharps exalted it in reverent terms. Confronting a ferocious bull, hunter John Cloud Jacobs remembered "praying to the hunting god—Mr. Sharps." Frank Mayer declared that "if my life depended on one shot from one rifle and I could take my choice, I'd rather have my old 'Sharps Buff' in my hands than any other gun."[24]

The early Sharps buffalo rifle was a true heavyweight, with models ranging from twelve to sixteen pounds. The bulk of this weight resided in a thick, octagonal barrel that could absorb the explosion of a 100-grain charge of black powder while spitting forth a 473-grain bullet. Though Sharps rifles were available in a range of calibers, the "Big 50" was the most famous. In 1874, Sharps issued a new buffalo gun, the "Sharps Old Reliable." It weighed sixteen pounds and fired a 550-grain slug using a 120-grain charge of powder. Frank Mayer called it the "rifle to end all rifles . . . and I knew when I read about it the first time, my life would be blighted until I owned one."

The blight on Mayer's life was short-lived. In 1875, he bought the new Sharps for $237.60, an amount he deemed a "small fortune." Hunter John Cook managed to purchase a Sharps, almost new, at the unheard of price of $36, including a reloading outfit, a bandoleer, and 150 cartridges. Cook explained that his good fortune came about because the man who sold him the rifle was getting out of the business, "having met with the misfortune of shooting himself seriously, but not fatally, in the right side with the same gun."[25]

Whatever they paid for their Sharps, the men who bought them thought it a bargain. When Mayer hit a buffalo with his Old Reliable, "I didn't have to inquire whether he was down for good." The Sharps's knockdown power was also delivered with remarkable accuracy, an attribute that could be enhanced with the purchase of added features such as telescopic sights and hair-triggers. With a German-made, twenty-power scope, Mayer claimed (credibly) to have killed 269 buffalo with 300 cartridges at a range of 300 yards.[26]

So legendary was the range and power of the Sharps that the Indians called it the gun that "shoots today, kills tomorrow." One of the greatest Indian defeats in Texas history came as a direct result of their underestimating the potency of the Sharps. In 1874, responding to the invasion of their land by buffalo hunters, a combined force of some 700 Comanche, Cheyenne, Kiowa, and Arapaho warriors attacked a crude outpost known as Adobe Walls. Only 29 whites occupied the post. Many of them, though, carried Sharps buffalo rifles. For three days the Indians laid siege, mounting charge after charge. Each time they were repulsed, the Sharps dealing out lethal force at great range. One Indian, reportedly, was shot from his horse at 1,500 yards.[27]

THE SHARPS BUFFALO RIFLE, TOGETHER WITH THE RAILROAD, BROUGHT the industrial revolution to the Great Plains. The Sharps's range, accuracy, and punch made it the perfect weapon for efficient, commercial slaughter—the mass production of hides.

In the early days of the hunt, one challenge the men did not face was finding their quarry. "It was easy enough to find the buffalo," remembered John Cloud Jacobs. "We would go a mile or so from camp to begin the day's hunt." For Jacobs, the ideal group would number between twenty and seventy animals, preferably close to cover such as a coulee or the crest of a butte.[28]

Before deciding on the direction from which to approach the herd, Jacobs checked the wind by dropping a few blades of grass. The buffalo relied on scent more than sight for protection, and hunters always approached from downwind. Still, at a distance of about 600 yards,

Jacobs began to worry about the sentinels that each herd maintained. He stooped over, and "[s]o long as our course was straight, up to a distance of about 400 yards, they could not make out what we were." At 400 yards, Jacobs dropped down and crawled, veering sideways only if forced by such obstacles as "a bunch of prickly pear or a stubborn, diamond-backed rattler that would not break ground." Most hunters sought a position around 300 yards from their quarry; any closer and the shooting might cause a stampede. From 300 yards, though, a skillful runner might make a stand—state-of-the-art industrial hunting.[29]

Gunning down a herd of buffalo without causing a stampede required good aim as well as keen insight to animal behavior. From a prone position on the plains floor, experienced runners took the time to study the herd, picking out the leaders and the sentinels. Ammunition was spread out for easy access, along with the second rifle and cleaning rods. Many hunters used shooting sticks, crossed pieces of wood on which they rested their gun barrel for a steady aim. Some achieved the same purpose with a stack of buffalo chips.

The Stand: A hunter finds the ideal condition for a stand in this painting,
The Still Hunt, by James Henry Moser.
Courtesy of the Jefferson National Expansion Museum, National Park Service.

The runner's first target was usually the old cow that typically led the herd. "Buffalo society was a matriarchy, and the cow was queen" was how Frank Mayer explained it. "Wherever she went, the others, including the big bulls who should have known better than follow a woman, went." The basic strategy for the stand centered on this lead cow. "When she got into trouble, they didn't know what to do. And our job as runners was to get her into trouble as soon as we could."[30] Vic Smith, who killed a record 107 buffalo in a single stand,[31] liked to put his first shot just in front of the lead cow's hip. Other hunters shot the leader through the lungs. The goal, in any case, was the same—not to kill the lead animal but to wound it. Not only was a wounded leader immobilized but she also drew the confused attention of the remaining herd. "Then the rest was easy," said Mayer.

Easy enough that an experienced hunter took his time, alternating rifles to keep the barrels from overheating, even cleaning his guns every five or six shots. Hunters differed in their preferred placement for the kill shots. Some aimed for the neck; others the heart. Runner John Cook put the sights of his Civil War Enfield on what he called the "regulation place," defined as "anywhere inside of a circle as large as a cowboy's hat, just back of the shoulderblade."[32]

John Cook killed eighty-eight buffalo in his best stand. As he poured shot after shot into the herd, Cook was amazed that "some would lie down apparently unconcerned about the destruction going on around them." At one point, the smoke from his rifles was so thick that Cook was forced to move to a more favorable position. The stand finally broke when he missed his "regulation place," shooting a big bull in the leg. The bull "commenced 'cavorting' around, jamming up against others, and the leg flopping as he hopped about." Then the bull somehow managed to run off, pulling the remaining survivors in tow.[33]

Barring a misplaced shot, the number of buffalo that a skilled runner could harvest in a stand was limited primarily by the capabilities of his skinners. "We never killed all the buff we could get," remembered Frank Mayer, "but only as many as our skinners could handle." In his brutally blunt fashion, Mayer explained the governing economics: "Killing more than we could use would waste buff, which wasn't important; it also would waste ammunition, which was."[34]

A mountain of hides awaits transport east at Wright & Rath's Hideyard
in Dodge City, Kansas.
Courtesy of the Kansas State Historical Society.

"TOUCH A SKINNING KNIFE?" THE QUESTION, FOR FRANK MAYER, was
merely rhetorical. "Not on your life! We had a caste on the buffalo
ranges, and I wouldn't put a skinning knife in my hand. I was exalted, a
runner, not a skinner." Skinning, by general consensus, was a "dirty, disagreeable, laborious, uninspiring job." Someone, though, had to do it,
and in the tough economic climate following the Panic of 1873, a wage
of 25 cents a hide was enough to make plenty of men willing. A good
skinner could strip 20 to 40 hides a day—the record was 70. The army,
for comparison, paid privates $13 a month.[35]

Buffalo skinners were easily identifiable by their filthy clothing—stiff
with dried blood—and by their stench. The skinner was a "paradise for
hordes of nameless parasites." Western saloons would not admit skinners

until they bathed. The frontier, no showcase for personal hygiene, couldn't stomach the reek of skinners.[36]

John Cook was one of the few men who violated the caste of the buffalo range, working simultaneously as both hunter and skinner. Wyatt Earp, who later won fame at the OK Corral, was another. Cook had a vivid memory of his first day of skinning. "I would not attempt to tell the different positions and attitudes I placed myself in that day." After skinning ten buffalo, "I was so stiff and sore I could hardly get out of the wagon."[37]

It was gory, stinking work. On cold days, skinners worked fast to keep the hides from freezing inseparably to the flesh, periodically shoving their hands into steaming entrails to save their fingers from frostbite. In the heat, skinners battled swarms of flies as they hurried to complete their work before the carcasses bloated, which also made it impossible to remove the skin. Summer hides were sprinkled with powdered arsenic to kill the fleas.[38]

Some skinners hitched up a team of horses to rip the hide from the carcass, hoping to boost efficiency. The hides, indeed, came off more quickly. Unfortunately, they often came off in more than one piece, ruining their value.[39]

Once a hide was removed, skinners cut holes around its perimeter and staked it out, hair side down, on the floor of plains. A hide dried out in a little over a week. In the old days of the Indian trade, these flint hides were then worked into supple robes, a laborious process usually undertaken by women. In the industrial age, though, flint hides were simply rolled or folded into bales. Hunters transported these bales to railhead processing centers such as Denver, Leavenworth, and Fort Worth. John Cook remembered the common sight of caravans, twenty-five wagons in length, each buckboard piled with 100 to 200 hides. At the railheads, huge warehouses were constructed to store the hides before shipment to tanneries in Chicago, Kansas City, and Europe. In Dodge City, the railside hideyard of Wright & Rath was sometimes stacked with 70,000 hides at a time. Between 1872 and 1874, an estimated 850,000 hides transited Dodge alone.[40]

Industrial processing of flint hides even changed the seasons in which buffalo could be hunted. For the robe trade, only the plush hides of winter sufficed. But with the new techniques for turning buffalo hides into leather, summer hides worked just as well. Hide hunting became a year-round enterprise.

FOR ALL ITS EMPHASIS ON EFFICIENCY, THE INDUSTRY OF HIDE HUNTING was synonymous with waste.

Most notorious was the market hunters' almost complete disregard for the meat. Early buffalo hunters, such as Bill Cody and J. Wright Mooar, had hunted for the meat and not the hide. Much of this fresh product stayed in the West, feeding railroad crews and frontier towns. Choice cuts could also find a market in the East. Tongue was considered a delicacy and could easily be transported when salted and packed in barrels. Mooar and other meat hunters also processed buffalo tenderloins and "hams" by curing the meat in brine and then sewing it into tight canvas wrapping. For decades, a buffalo ham was an American holiday tradition.[41]

Mooar was one of the few hunters to harvest meat throughout his buffalo-hunting career. For his trouble, he received 2.5 or 3 cents per pound, sometimes giving up half of that amount to pay the freighter.[42] For most hide hunters, the trouble and expense of harvesting meat weren't worth it. As the herd moved farther and farther away from settlements and the railroad, transport of fresh meat became increasingly difficult and often impossible. Nor, in the hide-hunting era, did it make economic sense to spend the time required to cure choice cuts. A man could earn far more by simply moving quickly to the next $3 hide. The average hide hunter harvested only the tongue, hacked out from below the jaw. A tongue added an easy two bits to each kill.

As for the rest of the carcass, it rotted on the plains—400 to 600 pounds of meat per animal, *millions of tons* of meat, even as famine gripped Indian tribes and the poor in eastern cities. Frank Mayer remembered that "the stench was so great that at a mile away from a stand you could smell it and be forced to hold your nose." One hunter described the carcasses on the plains so thick "that they would look like logs where a hurricane had passed through a forest."[43]

The only beneficiaries of this waste were the wolves and the coyotes. "To them it was a field day," said Mayer. "[A]ll the rotten meat you wanted to eat without the necessity of hunting it and running it down. What could be finer, from a coyote's point of view?" The population of

Industrial revolution on the plains: A period sketch shows various steps
in the processing and transport of hides. Note also the pile of bones.
From Harper's *magazine.*

wild canines exploded, spawning in turn the "wolfer"—the lowest caste
in the buffalo country hierarchy. Mayer, no prissy, called wolfers "a mean,
ugly, cheap breed of drunkards mostly." Wolfers followed in the destructive wake of the buffalo runners, sprinkling carcasses with strychnine,
then sitting back to wait. Wolf hides sold for a buck.[44]

More surprising than the buffalo hunters' well-known waste of meat
was the lesser-known waste *of hides.* One reason that "one-shot kills"
were important was that peppering a hide reduced its value on the market. So did sloppy skinning. So could untimely rain or infestations of
beetles. John Cook estimated that for every hide that he sold, one and a
half were unsalable. His outfit was relatively proficient; less skillful crews
might require four kills to get a single hide to market.[45]

BY THE END OF 1873, THE BUFFALO WERE GONE FROM KANSAS.

Some hunters professed shock. According to John Cook, "it was the
opinion of conservative hunters as late as the New Year of 1877 that the

present army of hunters were not killing the original herds, but only the natural increase." One runner lamented that he and his partner had "expected to be buffalo hunters all our lives." (They turned instead to ranching.)[46]

If they were surprised, the commercial hunters had overlooked ominous signs. Most obvious, of course, was the fact that the buffalo, once everywhere, became suddenly difficult to find. "I found it out every day when I went out scouting for something to shoot," recalled the clear-eyed Frank Mayer. "A couple of years before it was nothing to see 5,000, 10,000 buff in a day's ride. Now if I saw 50 I was lucky. Presently all I saw was rotting red carcasses or bleaching white bones." Less obvious, but eerily foreboding, were reports of herds composed entirely of calves.[47]

Those who understood the signs ignored them. Mayer, nothing if not honest, would remember that "[i]t wasn't long after I got into the game that I began to realize that the end for the buffalo was in sight. I resolved to get my share." Hunter Charles "Buffalo" Jones, who would later see the slaughter of the buffalo from a very different vantage, explained the enticing logic that fueled the continuing carnage.

> *Often while hunting these animals as a business, I fully realized*
> *the cruelty of slaying the poor creatures. Many times did I*
> *"swear off," and fully determine I would break my gun over a*
> *wagon-wheel when I arrived at camp.... The next morning I*
> *would hear the guns of other hunters booming in all directions*
> *and would make up my mind that even if I did not kill any*
> *more, the buffalo would soon all be slain just the same.*[48]

John Cook, like Buffalo Jones, would write in his memoirs about the regret he felt in the aftermath of a stand. "As I walked back through where the carcasses lay the thickest, I could not help but think that I had done wrong to make such a slaughter for the hides alone." Cook would also describe, without apparent irony, a song that he loved to hear. "The words were sung all over the range . . . and with the melody from the hunters' voices it was beautiful and soul-inspiring to me." He could not recall the song's title, but he remembered its first line: "O, give me a home, where the buffalo roam . . ."[49]

THE CARNAGE OF 1872 AND 1873 CAUSED GREAT ALARM AMONG THE INDIAN tribes who for centuries had hunted buffalo on the Kansas plains. Their hopes for preserving their way of life depended on an 1867 treaty with the United States. The Medicine Lodge Treaty gave tribes of the Southern Plains (Kiowa, Comanche, Cheyenne, and Arapaho) exclusive access to a large reserve that covered parts of southern Kansas, Oklahoma, and the Texas Panhandle.

The terms of the treaty seemed unambiguous. In exchange for granting the United States the land it needed to build the Kansas Pacific Railroad, the tribal lands were to be "set apart for the absolute and undisturbed use of the Indians." No person besides the Indians and Indian agents was allowed to "pass over, settle upon, or reside in the territory."[50]

The clarity of these terms, however, was about to confront a force more powerful than law. J. Wright Mooar, the young man from Vermont, would again find himself at the vanguard. As the fate of the Kansas buffalo began to become obvious, Mooar and another hunter undertook a scouting trip in an effort to discover the next great hunting ground. They traveled south and west until, in the Panhandle of Texas, they found their prey, "millions upon millions." "For five days," remembered Mooar, "we had ridden through and camped in a mobile sea of living buffalo."

But as Mooar was well aware, there was an irksome hitch. Mooar's "sea of living buffalo" lay squarely within the lands set aside for Indians by the terms of the Medicine Lodge Treaty. Back in Kansas to report his findings, Mooar and his fellow hunters "held a council" to determine their next step. Ultimately, they decided to seek official advice, selecting Mooar and a man named Frazier to act as their envoys to Fort Dodge.

The major in command of Fort Dodge, as Mooar described him, was a gracious host. They talked at length about the buffalo trade before Mooar finally got down to business. "Major, if we cross into Texas, what will be the Government's attitude toward us?"

The major's response was gracious indeed. "Boys," he declared, "if I were a buffalo hunter, I would hunt buffalo where the buffalo are."

"That," remembered Mooar, "settled the question."[51]

In 1873, the Kansas herd decimated, white hunters spilled south into Indian lands. A few army officers and white Indian agents attempted to turn them back but relented in the face of antipathy, including threats on their lives. The aggrieved tribes took their complaints all the way to President Grant, who agreed that buffalo hunters were violating the Medicine Lodge Treaty. The commander in chief, however, issued no order to force the whites out.[52] On the front lines, it was "hunt buffalo where the buffalo are" that ruled the land.

Inevitably, fighting broke out between white hunters and the Indians, including the three-day Battle of Adobe Walls. That battle, according to Texas historian T.R. Fehrenbach, would precipitate "the last great Indian war in Texas."[53]

Harlan B. Kauffman was one of the salty buffalo hunters whose Sharps rifle repelled the Indians at Adobe Walls. As Kauffman saw it, "the passing of both the Indian and the buffalo was inevitable. The great development of the West could never have begun until their occupancy ended." John Cloud Jacobs agreed:

> With the end of the buffalo the Indian depredations were over. They lived on the buffalo, and came in and murdered our women and children. After the buffalo were gone, the Government had no trouble in keeping them on their reservations, and the range was soon settled by thrifty farmers and ranchmen.[54]

Wyatt Earp was another buffalo hunter who articulated the prevailing viewpoint: "An examination of the record will show that, as a whole, the bad Indians were subdued into good Indians almost concurrently with the slaughter of the buffaloes, their one source of livelihood on the open range." Earp offered no opinion on the righteousness of this fact. "Whether the means was justified by the ends is not for me to say."[55]

Economics alone would have been wholly sufficient to doom the buffalo. Certainly the Kansas plains of 1872 and 1873 demonstrated the ruthlessly brutal power of this incentive. In the remainder of the decade, however, a powerful *strategic* rationale would also come into play: that the destruction of the buffalo was necessary if the nation were to prevail in the intensifying war with the Indian. As with so many landmark events in the nineteenth century, Grinnell would be there to watch it happen, to participate in the history unfurling on all sides around him.

CHAPTER SIX

"That Will Mean an Indian War"

Ross: "There'll be lots of men in here this winter, I guess, but if not this winter, surely next spring."

Grinnell: "Well, that will mean an Indian war I expect, for the Sioux and Cheyennes won't give up this country without a fight."

JULY 26, 1874[1]

By the spring of 1874, George Bird Grinnell's father had extricated the family enterprise from the vicious grip of the Panic of 1873. Twenty-four-year-old Grinnell, meanwhile, had extricated himself from the family enterprise. Under the tutelage of his mentor, Professor Marsh, Grinnell began pursuit of a doctoral degree from Yale, while also working as an assistant for paleontology at the new Peabody Museum.

Though his studies placed him at a quintessential East Coast establishment, Grinnell kept an eye to the West. Opportunity to return to the land that he loved was not long in arriving. In late spring, Professor Marsh received a letter from his friend, General Philip H. Sheridan, the Civil War hero then serving as commander of the U.S. Army's Division of the Missouri. Sheridan told Marsh that the army would send a large expedition to the Black Hills of the Dakotas that summer. As was common practice at the time, the army would take along several scientists. Sheridan inquired as to whether Marsh, who by now boasted a national reputation for his research in the West, would want to send along "a man to collect fossils." Marsh offered the position to Grinnell, who eagerly accepted. Grinnell, in turn, invited his friend Luther "Lute" North, the young scout who had guided him on the buffalo hunt with the Pawnee.[2]

By June Grinnell was on a train for Chicago, there meeting personally with General Sheridan and receiving all necessary paperwork. Then it was on to Fort Abraham Lincoln in the Dakota Territory, the staging ground for the expedition. On the train to Bismarck, the westernmost terminus of the bankrupt Northern Pacific Railroad, Grinnell met the man who would lead the Black Hills expedition—General George Armstrong Custer.[3] Over the next three months, Grinnell witnessed firsthand the events that would catalyze the final chapter in the American war against the Plains Indians.

A TRAGIC PATTERN DEFINED RELATIONS BETWEEN AMERICA AND ITS native tribes. As a growing nation expanded ever westward, Indians stood in the way. Conflict resulted, and as a consequence, the Indians were pushed onto smaller and smaller lands. As more and more Americans went west in the half century between 1840 and 1890, the country ran out of frontier, until there was no longer a place for the Indians and their traditional way of life.

Two consistent themes permeate the relationship between whites and Indians. First, when conflict arose between promises (that is, treaties) and economic expedients (travel routes, gold, buffalo hides, farmland), economics trumped. Second, the vicious cycle of violence proved far more powerful than the virtuous efforts of would-be peacemakers. Thus, as in the most pernicious conflicts of our modern day, extremists held the cards.

In the West, the pattern played out from the moment of acquisition—indeed it was part and parcel of the Louisiana Purchase. Jefferson viewed the seemingly infinite lands beyond the Mississippi as a potential solution to the vexing conflict between whites and Indians. He envisioned a "permanent Indian frontier" to which tribes of the East could be relocated. In 1830, President Andrew Jackson formalized this doctrine with his signature on the Indian Removal Act. Under this law, Indian title to eastern lands was extinguished and entire tribes were forcibly moved. Most infamous was the forced march of the Cherokee from their native lands in Georgia to a reserve in what is now Oklahoma. Of the 15,000 Cherokee forced from Georgia, 4,000 would die along a path now known as the Trail of Tears.[4]

As for the tribes already west of the Mississippi, the four decades following the Lewis and Clark expedition were relatively peaceful. The West, in this era, was still big enough to accommodate both the economic aspirations of whites and the traditional lifestyle of Native Americans. Certainly bloody confrontation took place, including, for example, almost constant warfare between fur trappers and the Blackfeet. A more common form of interaction, however, was trade, and far more Indians died from white disease than from white bullets. In 1837, for example, a smallpox epidemic reduced the Mandan tribe from approximately 1,600 to about 30.[5]

By the end of the 1840s, the swelling wave of emigration put whites and Indians on a new path to confrontation. The caravans heading west along the California–Oregon Trail drove game away from traditional hunting grounds (ultimately dividing the buffalo into northern and southern herds). Groups of Indians, meanwhile, harassed the pioneers along the trail. In the 1851 Treaty of Fort Laramie, the United States attempted to clear a pathway by organizing the tribes into large, defined territories. The Sioux, for example, received an expanse of land encompassing large portions of what is today eastern Wyoming, southeastern Montana, and the Dakotas. In exchange for allowing the passage of emigrants and the building of forts along the California–Oregon travel corridors, the Indians were to receive an annuity of trading goods. In 1853, a similar treaty was negotiated with the tribes of the Southern Plains.[6]

The treaties of 1851 and 1853 failed both in keeping the whites and Indians apart and in keeping the Indians from their historical warfare with each other. As more and more whites flooded west, warfare broke out from the Pacific Northwest to the Texas plains and virtually everywhere in between.

Not even the Civil War stemmed the tide. In fact, the 1862 discovery of gold in Montana would lead to intensified fighting over the Bozeman Trail, a shortcut taken by miners between Fort Laramie and the goldfields. The path would soon be known as the Bloody Bozeman, for it ran directly through land promised by treaty to the Sioux. The Sioux, led by the great chief Red Cloud and supported by junior chiefs, including a young medicine man named Crazy Horse, waged a relentless campaign against a string of new forts that the army built to protect the road. In the greatest battle of what came to be known as Red Cloud's War, Crazy Horse decoyed an arrogant captain named William Fetterman from the relative safety of Fort Phil Kearny. Awaiting the captain just over a ridgeline were more 1,500 Sioux warriors. On December 21, 1866, Crazy Horse and the Sioux killed Fetterman and all eighty of his men in fighting that lasted less than an hour.

The shock of Red Cloud's War, a string of military debacles in Kansas, and a post–Civil War desire to keep the army small all helped to convince the U.S. government that a military approach to Indian affairs was not working. At the same time, the drive to advance the nation's grand obsession—the transcontinental railroad—put a premium on peace and stability. In this environment, the government found prudence to be the better part of valor, and it sued for peace. The result was two treaties. To make peace with the Indians of the Southern Plains, the United States signed the Medicine Lodge Treaty of 1867. To make peace with Red Cloud and the Northern Sioux, the U.S. entered the Fort Laramie Treaty of 1868.[7]

Red Cloud's War was arguably the most successful Indian campaign in history, and the Fort Laramie Treaty "ended the Red Cloud War on Red Cloud's terms." The Bozeman Trail was closed and the forts along it abandoned. A Sioux reservation was defined to include an enormous tract of land covering large parts of present-day Montana, Wyoming, Nebraska, and the Dakotas. As with the Medicine Lodge Treaty, the provision regarding trespass by whites was unambiguous: No one (except U.S. government employees) "shall ever be permitted to pass over, settle upon, or reside in the territory." At the core of the Sioux reservation was an area that the Sioux considered sacred—the Black Hills.[8]

The open question, of course, was whether the United States, having pledged its honor, would live up to the terms of its newest treaties. The answer, consistent with the historical pattern, was yes—until the higher cards of economics and politics came into play.

By the summer of 1874, the Medicine Lodge Treaty was already in shambles. The influx of Kansas buffalo hunters into treaty lands in Texas had sparked renewed warfare, and the tribes of the Southern Plains—the Comanches, the Kiowa, the Southern Cheyenne—were on the run. By the spring of 1875 it would be over, with all of these tribes forced forever onto greatly diminished reservations.

On the Northern Plains, though, Red Cloud's terms still prevailed. While there had been significant clashes between whites and Indians, the core terms of the Fort Laramie Treaty of 1868 remained in effect. Pressure on that treaty, however, was about to grow intense. Officially, Custer's mission in the summer of 1874 was to locate an appropriate site for a new fort. But hundreds of thousands of Americans knew there was a more intriguing reason for the ambitious general's exploration of the Black Hills.[9] The nation waited breathlessly for word of the expedition's results—because they knew Custer was hunting for gold.

ON A SUNNY JULY 2, THE GRAND PARADE THAT WAS THE EXPEDITION of 1874 rode out of Fort Abraham Lincoln. Sixteen musicians mounted on matching white horses set the step, playing the jaunty strains of "Garry Owen" and, with a chivalrous tip of the hat to Mrs. Custer, "The Girl I Left Behind Me." Thirty-four-year-old Custer, with keen instinct for the value of military pomp, cut a dashing figure at the head of one of the largest and best-equipped columns ever assembled in the West. He wore a buckskin shirt, a gray felt campaign hat, and a trademark red bandana at his neck. The general's ever-present hunting dogs—greyhounds and Scottish deerhounds—ran free, circling excitedly around his big bay horse.

Trailing behind the general were more than a thousand men, including ten companies of cavalry and two of infantry. There were 30 Arikara and Crow scouts, along with some 50 uniformed teens from the Santee Sioux Indian School. Supplies for this mass of men, including 100 tons of corn for the livestock, were portaged in 110 wagons pulled by six mules each. At the rear of the procession walked food for the soldiers—a herd of 300 cattle.[10]

In his three weeks at Fort Abraham Lincoln before the expedition's departure, Grinnell had occupied himself with hunting and the "collecting of bird skins." He also met the colorful cast of characters with whom he would explore the Black Hills. Foremost, of course, was Custer himself. Grinnell sat through several dinners at Custer's house, "where the General delighted to relate his hunting exploits, and where Mrs. Custer was extremely kind and hospitable."[11]

One of the civilians that Grinnell befriended was Charlie Reynolds, a man who might have stepped from the pages of Grinnell's boyhood dime novels. Reynolds already was widely respected among the frontier scout fraternity, and his exploits during the expedition of 1874 would vault him into the pantheon. There were other wild characters as well, including an Arikara scout named Bear's Ears. In his memoirs, Grinnell would record Bear's Ears's campfire story—a tale of losing the woman he loved to "an older and more important man." When Bear's Ears tried to kill his rival, he was banished from the tribe, then devoted the next several years to the cause of revenge. As a "sacrifice to the powers," Bear's Ears hacked off three fingers from his left hand. He ultimately caught up with his rival, killed him, and, in Grinnell's words, "gratified his revenge by eating a portion of the enemy's heart."[12]

Among the soldiers with Custer was a pensive engineer named William Ludlow, who would later play an important role in the preservation of Yellowstone National Park. Less pensive, perhaps, was Major Fred D. Grant, son of the president. Like his father, young Fred enjoyed an occasional nip at the bottle. After one binge during the expedition, Custer had Grant arrested.[13]

From the standpoint of history, no member of the expedition was more important than Horatio Nelson Ross, whom Grinnell described as "an old miner and prospector." While the army had essentially acquiesced to the presence of miners in the expedition, Custer himself was the driving force behind the search for gold. Custer had gone to the trouble of telegraphing his superiors to obtain specific permission for "the service of a geologist." The army agreed but wouldn't pay. Ultimately, the Minnesota government funded the participation of their own state geologist, a university professor named Newton H. Winchell. As for Ross, and another miner named McKay, one account has Custer funding them himself.[14]

RUMORS OF GOLD HAD BEEN PART OF BLACK HILLS LORE FOR DECADES, though as Grinnell remembered, the territory in 1874 remained "unknown—a region of mystery." One explanation for this continuing mystery—at a time when virtually all the surrounding territory had been explored—was the ferocity with which the Sioux had repelled white intruders on their holy ground.

Both the lore and the danger of the Black Hills are on display in the tale of a prospector named Ezra Kind. According to a story with at least some evidence to support it, Kind and seven partners departed Fort Laramie in 1833 and spent a year prospecting in the hills. The men found gold and, apparently, the Sioux. No member of the party was ever seen again, but a piece of sandstone was later discovered on the trail to Deadwood. On the stone was carved the following inscription, signed by Kind: "Got all of the gold we could carry our ponys all got by the Indians I have lost my gun and nothing to eat and Indians hunting me."[15]

Whether gold's existence was proven or not, the white citizenry of the Dakota Territory believed firmly in its presence in the Black Hills. In the decade preceding 1874, the cries for government exploration of the hills had been steady and growing. In 1865, the territorial legislature had formally requested a "geological survey." In Yankton and other towns on the path to potential goldfields, newspapers and civic groups kept up the drumbeat. (There was nothing like a gold rush to stimulate the local economy.) When the Fort Laramie Treaty of 1868 seemed clearly to forbid entrée to the Black Hills, these groups reacted with outrage. Yankton's *Press and Dakotaian* offered the following perspective:

> *This abominable compact with the marauding bands that*
> *regularly make war on the whites in the summer and live on*
> *government bounty all winter, is now pleaded as a barrier to the*
> *improvement and development of one of the richest and most*
> *fertile sections in America. What shall be done with these Indian*
> *dogs in our manger? They will not dig the gold or let others do*
> *it . . . They are too lazy and too much like mere animals to*
> *cultivate the fertile soil, mine the coal, develop the salt mines,*
> *bore the petroleum wells, or wash the gold.*[16]

The nationwide depression following the Panic of 1873 only added to disdain for the notion that valuable resources would be left untapped.

If Custer were eager to determine definitively the presence or absence of Black Hills gold, he stood on a solid foundation of public opinion (at least in the West). In fact, the notion of exploring the hills and *not* looking for gold could scarcely be entertained. Searching for gold would also ensure that Custer and his mission would receive something of which the general was quite fond—the attention of the public and press. Indeed, to ensure their firsthand knowledge of his exploits, Custer embedded in his expedition a number of reporters.

IN HIS MEMOIRS, GRINNELL WROTE THAT "GEN. CUSTER WAS A MAN of great energy. Reveille was usually sounded at 4 o'clock; breakfast was ready at 4:30; and by five tents were down and wagons packed, and the command was on the march." More than 300 miles of desolate prairie separated Fort Abraham Lincoln from the Black Hills, and in the flat, open country, the wagons rode four abreast. Even so, between the scouts ranging far out in front and the livestock bringing up the rear, the long column often stretched for miles. Grinnell, when not exploring on his own, was given the privilege of riding in the vanguard with Custer and his entourage. They averaged between fourteen and thirty miles a day, measured by a two-wheeled odometer, and Grinnell recorded that "camp was not reached until 8 o'clock at night, sometimes not till midnight, and sometimes not at all."[17]

Custer accorded Grinnell considerable latitude in his daily movements, and the young scientist (assisted by Lute North) pursued his official responsibilities with dogged determination. The pace of the expedition, however, allowed time for no more than superficial excavation. William Curtis, the correspondent for Chicago's *Inter Ocean*, wrote of Grinnell's toil that "it seems that nothing ever died in this region."[18]

Grinnell would be disappointed with his results as a paleontologist, though his discoveries did include two marine species new to science. General Custer himself took an interest in the young scientist's work, waxing enthusiastic in an official dispatch on July 14. "Mr. Grinnell of Yale College one of the Geologists accompanying the expedition discovered on yesterday an important fossil, it was a bone about four feet long and twelve inches in diameter and had evidently belonged to an animal larger than an elephant."[19]

In Grinnell's own writing, his discussion of the Black Hills expedition barely mentions dinosaur hunting. As with the Marsh expedition of 1870, it was hunting with a Winchester that sparked his greatest enthusiasm. Three hundred cattle, as it turned out, didn't feed an army for long. Fortunately, the country they traversed was thick with antelope and deer. Grinnell spent many of his days hunting with Lute North and scout Charlie Reynolds, men he clearly idolized. "The company never suffered for lack of fresh meat," remembered Grinnell, and "on several days the command had killed 100 deer or more."[20]

Custer too found time for the chase. Indeed it was the context of hunting that seems to have shaped some of Grinnell's sharpest impressions of the general. "General Custer was friendly, sociable and agreeable. He was very fond of hunting and a great believer in his skill as a rifle shot." As for Grinnell's assessment of Custer's ability, he noted dryly that the general "did no shooting that was notable."

Indeed Grinnell seemed to take a certain degree of pleasure in relating an incident in which Custer missed three consecutive shots at (literally) sitting ducks after boasting, "I will knock the heads off a few of them." Nor did Grinnell give credence to the general's claim that his greyhounds could run down antelope—the fastest land animal in North America. The dogs "did nothing of this kind on the Black Hills expedition," wrote Grinnell, though they "did kill plenty of jack rabbits." In a letter he wrote fifty years later, Grinnell described the general as giving "the impression always of being intensely in himself, somewhat egotistical in fact."[21]

Grinnell would also relate a story that reads, in hindsight, as ironic prophecy. The men of the Black Hills expedition might have seen an abundance of wild game, but they saw very few of the region's other indigenous inhabitants—the Sioux. Custer's massive army was easy to avoid, and deterring confrontation seems to have been a goal. The expedition did come across the remnants of what appeared to have been a

Custer the mighty hunter during the Black Hills expedition of 1874.
The man on the right is Colonel Ludlow, who in 1875 would lead
an expedition into Yellowstone that also included Grinnell.
Courtesy of the South Dakota State Historical Society–State Archives.

massive village. In campfire conversation that night, Lute North expressed a certain relief that contact had been avoided with so large a group of Indians. Custer's response: "I could whip all the Indians in the northwest with the Seventh Cavalry." Though he could not have known it, the results of the Black Hills expedition would set in motion a two-year chain of events leading directly to the valley of the Little Bighorn, where Custer would have the opportunity to put his bold prediction to the test.

THERE ARE CONFLICTING REPORTS OVER WHEN AND WHERE THE Custer expedition first discovered Black Hills gold. What seems clear from detailed journals kept by George Bird Grinnell is that he was among a small group of civilians who first heard the news.

According to Grinnell, on the night of July 26, 1874, he found himself in conversation with prospector Horatio Ross, Charlie Reynolds, and Lute North. Reynolds commented on the beauty of the Black Hills, a lush garden in comparison with the dry plains surrounding them.

"I don't wonder the Indians hate to have the white man come in," said Grinnell, "for of course if anybody should come in here and settle, the game would all be driven off pretty quick and like enough the hills burnt off too."

It was then that Ross shared his discovery. "I reckon the white people are coming all right." Turning to Reynolds, he asked, "Shall I show 'em that little bottle of mine, Charlie?"

"Of course," said the scout.

Ross turned to Grinnell and North. "Well now, boys, don't say anything about this, but look here."

Grinnell described how Ross then

> *drew from his pocket a little vial which he passed over to me and which we both examined. It was full of small grains of yellow metal which we of course knew must be gold dust. Some of it was very fine, but occasionally there would be nuggets as large as the head of a pin and a few as large as half a grain of rice. We turned the bottle round and round and by the light of the fire looked at the shining yellow mass which did not much more than cover the bottom of the vial.*

"That's gold, I suppose," said Lute North.

"Yes," replied Ross, "that's gold, and found right on the surface, and where there is as much as that, there's bound to be lots more when we get down a little lower. . . . There'll be lots of men in here this winter, I guess, but if not this winter, surely next spring."

The mood was surprisingly somber, perhaps because everyone knew the certain consequences of a gold rush in the Black Hills. "Well, that will mean an Indian war I expect," said Grinnell, "for the Sioux and Cheyennes won't give up this country without a fight."

The campfire cluster began to break up, anticipating an early morning. Lute North addressed himself to Ross before retiring. "Well, if you were working for yourselves, I'd wish you good luck, but as it is, I rather believe that I hope you won't find anything."[22]

In the coming days, though, Ross would find a lot (or at least enough). From August 1 through 5, Custer's column remained in "permanent" camp along the verdant banks of French Creek (near present-day Custer, South Dakota). Ross's secret was soon out, and according to reporter William Curtis, his early find "was tried with all the tests that the imagination of 1,500 excited campaigners could suggest and it stood every one. It was washed with acid, mixed with mercury, cut, chewed, and tasted till everybody was convinced and went to bed dreaming of the wealth of Croesus."

What awoke in the morning was an army of prospectors. Again from Curtis: "At daybreak there was a crowd around the diggings with every conceivable accoutrement: shovels and spades, picks, axes, tent pins, pot hooks, bowie knives, mess pans, kettles, plates, platters, tin cups, and everything within reach that could either lift dirt or hold it, and put into service by the worshippers of that God gold, and those were few who did not get a showing."[23]

While still in camp, Horatio Ross and twenty-one other members of the Custer expedition formed "District No. 1 Custer Park Mining Company, Custer's Gulch." Participants in the company, including Sarah Campbell, a black woman working as the sutler's cook, claimed mining rights to 4,000 feet of creek. To serve notice, they placed a written copy of their claim inside an empty box of hard tack and left it by the stream. Ross planned to return later that year, as did, no doubt, a number of

Custer's men. Desertions would soon spike. For an army private, his $13 a month in wages compared poorly against the wealth projected in those heady days on the banks of French Creek.

Custer, aware that the nation was watching, did not make it wait. On August 2, 1874, he sent Charlie Reynolds and a saddlebag full of dispatches on a dangerous solo journey to Fort Laramie, then about ninety miles away. In Elizabeth Custer's widely read autobiography, *Boots and Saddles*, she would claim that Reynolds arrived at Fort Laramie with his lips "parched and his throat so swollen that he could not close his mouth."[24] Given the fort's proximity to the North Platte and Laramie Rivers, it seems unlikely that Reynolds arrived thirsty. Nevertheless, the news he carried in a four-day journey through hostile country would seal his legend.

In fairness to Custer, his characterization of what he found was relatively circumspect—at least initially. The dispatch to General Sheridan carried by Reynolds reported that "gold has been found in several places, and it is the belief of those who are giving their attention to this subject that it will be found in paying quantities. I have on my table 40 or 50 small particles of pure gold, in size averaging that of a small pin head, and most of it obtained today from one panful of earth." But, he added, "until further examination is made regarding the richness of gold, no opinion should be formed."[25]

Perhaps the caveat was thrown in to hedge his bets. By the time he wrote a dispatch dated August 15, Custer was losing a bit of his temperance. "[A] hole was dug 8 feet in depth. The miners report that they found gold among the roots of the grass, and, from that point to the lowest point reached, gold was found in paying quantities." Moreover, added Custer, "it has not required an expert to find gold in the Black Hills, as men without former experience in mining have discovered it at an expense of little time or labor."[26] Music to the ears, no doubt, of any man mired in the Panic of 1873.

The Yankton *Press and Dakotaian* had the news by August 13. "Struck It at Last!" Other headlines trumpeted, "Rich Mines of Gold and Silver Reported Found by Custer," "Prepare for Lively Times!" and "The National Debt to be Paid When Custer Returns."[27] It was a booster's dream.

William Curtis's reporting from the field would not reach Chicago until August 26. When it arrived, though, the editors of the *Inter Ocean* welcomed it with the screaming headline, "GOLD!" Little more need have been said, though Curtis added plenty. "They call it a $10 diggin's, and all the camp is aglow with the gold fever." By September 12, *Harper's Weekly* had upped the hype to "gold equivalent to $50 a day to the man if the yield should prove as good as promised."[28]

The first wave of the Black Hills gold rush arrived at French Creek almost before Custer made it back to Fort Abraham Lincoln. Indeed the "Collins-Russell party" began its excavations where Custer left off—in the bottom of one of Ross's holes. (Apparently, they either ignored or failed to notice the hardtack box containing the claim of "District No. 1 Custer Park Mining Company.") After confirming the presence of gold, according to a member of the party, "we set to work to fortify ourselves from attack by building a stockade."[29]

Once again, the fly in the ointment was a treaty with the Indians. Like the buffalo country of the Texas panhandle, the gold fields of the Black Hills lay squarely in Indian lands, and not even the most agile lawyer could construe it otherwise. For a while, the army engaged in a halfhearted effort to keep the argonauts out of the hills. As would soon become obvious, however, the real goal of the U.S. government was not to convince the whites to stay out but rather to convince the Indians to let the whites in.

The most straightforward solution was simply to buy the land from the Sioux. In the spring and fall of 1875, the government called Red Cloud and other Sioux leaders to Washington, D.C., for negotiations. Offered $6 million at one point, the Sioux refused to sell. Frustrated, the commissioners charged with representing the United States would recommend that the government simply fix a price for the land it wanted, then demand that the Indians accept it.[30]

Negotiations, in any case, would soon be overtaken by events. In December 1875, the army issued an ultimatum to the far-flung Sioux nation: Report to agency officials by January 31, 1876, or be considered hostile. Far-flung bands of Sioux were spread from Nebraska to Montana, and some groups likely never received the message. Others ignored it, perhaps having come to the realization that there could be no more retreat.

In the spring and summer of 1876, defiant tribes of Sioux and Cheyenne had begun to congregate on the plains of eastern Montana. Their leaders included Sitting Bull, Crazy Horse, and Gall. Ultimately they would form a massive single village, perhaps 7,000 strong, strung out along a shallow creek called the Little Bighorn.

On May 17, 1876, General George Armstrong Custer led another procession out of Fort Abraham Lincoln. The expedition of 1876 traced its lineage directly to the Black Hills venture of two years before. On the banks of French Creek, as Custer would soon discover, he had dredged up a good deal more than just gold.

THE BUFFALO PLAYED A CENTRAL ROLE IN THE DECADES-LONG WAR between the United States and the Plains Indians. The influx of emigrants along the central plains travel corridors caused early friction, in part because of its impact on the buffalo and traditional hunting grounds. Eventually, the invasion of Texas reservation lands by commercial buffalo hunters would spark the final chapter in the war against the Indians of the Southern Plains.

Destruction of the buffalo was not only a cause of warfare but also a weapon of warfare—total warfare against the Indians. As the U.S. Army became increasingly frustrated at its inability to gain the upper hand in the guerrilla-style war, growing attention focused on the significance of the Indians' food supply. Certainly the reliance of the Indians on the buffalo was obvious to all. Theodore R. Davis, who chronicled the "buffalo range" for the readers of *Harper's New Monthly Magazine* in 1869, wrote that "if, as the Indian fears—groundlessly, however, at present—the buffalo will pass away, I am at a loss to know what he would do, for the buffalo feeds, clothes, and warms the nomads."[31]

Though the United States never adopted an official policy of destroying the buffalo, the pervasiveness of the doctrine seems undeniable. The commanding general of the army during the finale of the Indian wars was William Tecumseh Sherman, notorious for his Civil War "march to the sea." In the South, Sherman conquered the Confederates with "total warfare," a strategy that sought to defeat the enemy not only militarily but also economically and even psychologically. What total warfare sought to destroy, ultimately, was the will to resist, and a fundamental tenet of the doctrine included an unflinching willingness to make the civilian population pay a price for the war. As Sherman's army marched from Atlanta to Savannah, they killed livestock, destroyed crops, and burned cities and farms.

In the West, Sherman would bring a penchant for the same scorched-earth strategy to the war against the Indian. His deputy was another Civil War hero, General Philip Sheridan, whose considerable reputation had been won in the application of the total-warfare doctrine to the valley of the Shenandoah.[32]

General Sheridan is widely attributed with making a speech about the buffalo that he probably never delivered. According to lore, Sheridan testified before the Texas legislature against an 1875 law that would have protected the buffalo: "Let them kill, skin and sell until the buffalo is exterminated, as it is the only way to bring a lasting peace and allow civilization to advance." Historians, though, can find no hard evidence to support the tale. (They *can* document other of Sheridan's utterances, including "The only good Indians that I ever saw was dead."[33])

The government's support for the destruction of the buffalo, in any event, did not require sweeping policy pronouncements. As buffalo hunter Frank Mayer explained, "Don't understand that any official action was taken in Washington and directives sent out to kill all the buff on the plains. Nothing like that happened. What did happen was that army officers in charge of plains operations encouraged the slaughter of buffalo in every possible way."[34]

Anecdotal support for Mayer's contention is plentiful. A Nebraska officer encouraged a traveling British gentleman to "kill every buffalo you can. Every buffalo dead is an Indian gone." From the Southern Plains had come the law-be-damned encouragement to "hunt the buffalo where the buffalo are." When Sitting Bull eventually retreated into Canada, the army burned thousands of acres of grass along the border to prevent the herd from moving north. One of the most practical forms of support, greatly appreciated by cost-conscious buffalo runners, was the army's distribution of free ammunition so long as a hunter swore to "shoot it into a buffalo."[35]

Some viewed destruction of the buffalo as a humanitarian alternative to the killing of Indians. "Either the buffalo or the Indian must go," said

one plains officer. "Only when the Indian becomes absolutely depen-
dent on us for his every need will we be able to handle him. He's too
independent with the buffalo. But if we kill the buffalo we conquer the
Indian. It seems more a humane thing to kill the buffalo than the Indian,
so the buffalo must go."[36]

Just as important as the military's support for the destruction of the
buffalo was the degree to which the Indian wars provided a rationalization among the civilians who carried out the carnage. John Cook
admitted that he sometimes felt a "slight feeling of remorse" for his role
in the "greatest of all 'hunts to the death.'" He took comfort, though, by
recalling his part in the great national enterprise of settling the West, of
transforming wildlands in the production of civilization:

> *I pictured to myself a white school-house on the knoll yonder*
> *where a mild maid was teaching future generals and statesmen*
> *the necessity of becoming familiar with the three R's. Back*
> *there on that plateau I could see the court-house of a thriving*
> *county seat. On ahead is a good site for a church of any Chris-*
> *tian denomination. Down there where those two ravines come*
> *together would be a good place for a country store and post*
> *office. Some of these days we will hear the whistle and shriek*
> *of a locomotive as she comes through the gap near the Double*
> *Mountain fork of the Brazos. And not long until we can hear in*
> *this great southwest the lowing of the kine, the bleating sheep,*
> *and the morning crow of the barnyard Chanticleer, instead of the*
> *blood-curdling war-whoop of the Kiowas and the hideous yell of*
> *the merciless Comanches.*[37]

J. Wright Mooar, pioneer of the buffalo industry, was more strident
still. He utterly rejected the criticism of the commercial hunter "as a
destroyer, a ruthless killer, a wastrel of great game resources." Were it not
for the hunters, he argued, "the wild bison would still graze where Ama-
rillo now is, and the red man would still reign supreme over the pampas
of the Panhandle of Texas." For Mooar, the balance was clear: "And I want
to state that any one of the families killed and homes destroyed by the
Indians would have been worth more to Texas and to civilization than all

the millions of buffalo that ever roamed from the Pecos River on the south to the Platte River on the north."[38] The buffalo hunters could not only look to their own financial well-being but could also do so knowing that their work contributed toward a broader national purpose.

By the mid-1870s the die was cast. Improvements in technology and infrastructure made wholesale slaughter feasible, while economic and military rationale provided twin-fisted incentive. The commercial buffalo hunter of the 1870s was well tuned to his gilded times. In a sense, what was remarkable was not the destruction of the buffalo but rather that someone emerging from the same era would stand up to oppose it.

CHAPTER SEVEN

"Ere Long Exterminated"

*It is certain that, unless in some way the destruction of
these animals can be checked, the large game still so
abundant in some localities will ere long be exterminated.*

—George Bird Grinnell,
*Report of a Reconnaissance to
Yellowstone National Park, 1875*[1]

I n the tug of war between Yale University and the West, Yale had a difficult time holding its grip on George Bird Grinnell. In the spring of 1875, less than a year after returning from his adventures with Custer, Grinnell received yet another irresistible invitation to adventure. This time the letter came from Colonel William Ludlow,[2] a military engineer whom Grinnell had come to know during the Black Hills expedition.

As Grinnell remembered it, "In the Spring of 1875 Col. Ludlow wrote me saying that in the summer he expected to be ordered to make a reconnaissance in Montana, and asking me to go with him as naturalist." Ludlow's specific destination would be the recently established Yellowstone National Park. In Grinnell's characteristically understated manner, he noted in his memoirs that "Prof. Marsh seemed to think it desirable that I should go."[3]

The national park was only three years old in 1875, and part of Ludlow's mission was to improve the nation's understanding of what exactly had been preserved.[4] While the park itself had not been known for Indian problems, the Ludlow expedition would have to traverse 300 miles of highly contested Montana ground to reach it. Ludlow would be accompanied by a detachment of cavalry and, to Grinnell's delight, his friend Charlie Reynolds would serve as civilian scout.

The association between Grinnell and national parks that began in the summer of 1875 would endure for the rest of his life. It was Yellowstone where the 26-year-old's environmental philosophy would begin to coalesce. And it was Yellowstone, ultimately, where Grinnell would fight the last stand to save the buffalo from extinction.

ON AUGUST 16, 1806, AS LEWIS AND CLARK DESCENDED THE MISSOURI River on their return from the Pacific, they released from military service one of the most important members of their expedition. His name was John Colter, and at a time, after three years in the wilderness, when most members of the Corps of Discovery were eager to go home, Colter's appetite for exploration was not yet sated. At the Mandan villages the corps met a party of trappers headed back up the Missouri. Colter wanted to join them, and William Clark would record in his journal that "we were disposed to be of Service to any one of our party who had performed their duty as well as Colter."[5]

Colter's first trapping foray was apparently unsuccessful, but by 1807 he was in the employ of Manuel Lisa, the great St. Louis fur trader. Lisa established a crude fort at the confluence of the Yellowstone and Bighorn rivers, hoping to trade with the Indians. In the dead of winter, Lisa dispatched Colter on a 400-mile snowshoe trek to establish relations with the surrounding tribes. It was on this journey that Colter entered history as the white discoverer of what is today Yellowstone National Park.

Colter himself left no written record of his feat. When he finally returned to St. Louis in 1810, however, he recounted his travels to William Clark, by then brigadier general of the militia and agent of Indian affairs for the territory of Missouri. Clark was at work on a map of the Lewis and Clark expedition. On the basis of the information he learned from Colter, Clark added a dotted line designated "Colter's Route in 1807." On Clark's map, the Colter path skirts the western shore of a large lake (then called Lake Eustis) near the headwaters of the Yellowstone River. Farther north along the Yellowstone, Colter's path is shown to ford the river at a point labeled "Hot Springs Brimstone."[6]

The motivation for Colter's 1807 exploration—the beaver trade—would be the same incentive that led other early white visitors to the upper Yellowstone. In 1827, a trapper named Daniel T. Potts wrote a lengthy letter to his brother (not made public until 1947) in which he described numerous details of Yellowstone, including "a number of hot and boiling springs some of water and others of most beautiful fine clay [that] resembles that of a mush pot and throws its particles to the immense height of from twenty to thirty feet."[7]

In the 1820s and 1830s, numerous mountain men traipsed the waterways of the Yellowstone and its myriad tributaries. Unlike Potts, most did not record the details of their journeys in writing. Instead, knowledge of the Yellowstone region spread via the trappers' great oral tradition. Part of this tradition included a well-known proclivity toward exaggeration, and for decades, owing to the source, the reports of Yellowstone's strange wonders were largely dismissed.

Jim Bridger, "King of the Mountain Men," was one early visitor to Yellowstone. His geographic knowledge of the region was sufficient to provide the basis for a detailed map drawn by Father Pierre-Jean DeSmet in 1851. But to Bridger, perhaps more than any other man, would be attributed a number of the tall tales that comprise Yellowstone's early lore. These include the "echo used as an alarm clock," the "fish cooked in hot water," and the "petrified trees a-growing with petrified birds on 'em a-singing petrified songs." Bridger himself actually propagated some of these tales, while others are merely attributed to him. In any event, the skeptical audience for Yellowstone stories is not hard to understand.[8]

Nor did the next wave of Yellowstone explorers, the prospectors, offer much of a reputation for veracity. In 1862, wandering argonauts discovered gold in a small Montana waterway known as Grasshopper Creek, some 100 miles west of Yellowstone. Much as trappers had done before, miners were soon probing every waterway in the region. Fortunately for the future of Yellowstone, gold was not among the treasures they found. Many prospectors' tales, though, reported otherwise. According to one story, which actually found a newspaper to print it, a party of five men exploring the headwaters of the Yellowstone stumbled on an enormous "Indian catacomb." Inside the tomb were the remains of

thirty warriors wearing golden bangles, and at the center of the crypt was a "massive basin or kettle" that "proved to be pure gold."[9]

If miners' stories did not meet the standards of credulity for nearby Montanans, they nevertheless succeeded in planting the seeds of considerable curiosity. What followed in 1869, 1870, and 1871 were a succession of three formal exploratory expeditions. Each fell like a domino on the next, culminating with the establishment of the world's first national park.

In the summer of 1869, Montana's territorial newspapers began to write about the possibility of an organized exploration of the upper Yellowstone by the army and a civilian coterie of "some of the most prominent men in the Territory." One man who hoped to go along was David Folsom, a transplanted Quaker from Maine who had dabbled in mining, hunting, ranching, and most recently—ditch digging. Folsom, though skeptical of miners generally, had noticed that "however much the accounts of different parties differed in detail, there was a marked coincidence in the descriptions of some of the most prominent features of the country."[10]

David Folsom, the Maine Quaker who led the first of three critical explorations of Yellowstone.
Courtesy of the National Park Service, Yellowstone National Park.

By the end of the summer, the proposed expedition appeared ready to disintegrate before a single mile had been logged. Some of those who had committed to the trip were prevented by "pressing business engagements." Then came news that the army would be unable to participate, a development that gave pause to many in light of recent Indian troubles. By September, only three stalwarts remained: David Folsom and two of his friends—Charles Cook and William Peterson. On September 6, 1869, the three men departed.

In a thirty-six-day exploration, the Folsom party documented most of the major Yellowstone landmarks viewed by today's tourists, including hot springs, mud pots, the Grand Canyon of the Yellowstone, the falls, Yellowstone Lake, and, in an area they called the Burnt Hole and Death Valley, "an intermittent geyser in active operation." The geyser, according to Folsom, "shot into the air at least eighty feet."[11]

On the return of the expedition to Helena, Folsom and Cook received their own taste of the skepticism greeting Yellowstone explorers. The influential *Scribner's Monthly* and *New York Tribune* both declined to publish a detailed account compiled from the two men's journals.[12]

One man who believed what he heard from Folsom was Nathaniel P. Langford, a former Montana tax collector, unemployed in the summer of 1870. Langford knew a little bit about taking the initiative, having earlier played a prominent role in organizing a force of Montana vigilantes who strung up a notorious gang of thieves (led by the local sheriff) in the boomtown of Bannack. On a trip back east, Langford met with another man of action, railroad financier Jay Cooke. Just three years later, the bankruptcy of Cooke's firm would spark the Panic of 1873. In 1870, though, Cooke was in full booster mode, searching out any angle to attract investment for construction of the Northern Pacific Railroad.

Thus began a long and multicolored relationship between the railroad and Yellowstone. Encouraged by Langford, Cooke believed that the wonders of the region could encourage tourism to Montana—and thereby help promote the construction of the Northern Pacific. With financial support from Cooke, Langford returned to Montana to help organize a new and larger exploration of Yellowstone.[13]

With Langford as the driving force, the expedition of 1870 now took shape. Its official leader would be Henry Washburn, a former Union

Army general and the surveyor-general of Montana. A number of other prominent Montanans and an escort of cavalry would also make the trip, lending the party a credibility that trappers and prospectors had lacked. The Langford–Washburn expedition spent more than three weeks in Yellowstone, confirming the presence of the incredible features long reported by earlier visitors.

This time, the wonders of the park would begin to seep into the broader national consciousness. Several expedition members published articles in prominent national magazines. *Scribner's Monthly*, for example, was now a believer, printing a lengthy piece by Langford. Langford, as part of his arrangement with Cooke, also launched a multicity lecture tour.

In January 1871, Langford spoke in Washington, D.C. One of the men in his audience was Ferdinand Hayden, the head of the U.S. geological survey of the territories. Inspired by Langford's lecture, Hayden sought funds from Congress for a large, federally funded exploration of Yellowstone. Congress put up $40,000, and in the summer of 1871, a large party led by Hayden conducted the expedition that would lead most directly to the birth of Yellowstone National Park.[14]

Like victory, the idea for the creation of the first national park had a thousand fathers, if all the claims were to be believed. One long-standing legend, now discredited, was that members of the Langford expedition hatched the idea of protecting Yellowstone as a national park while chatting round the campfire on the banks of the Firehole River.[15] Many others would also declare their paternity of the park, sometimes bitterly.

Whoever first spawned the idea, action to establish the park was swift after the return of the Hayden expedition. In December of 1871, Montana delegate William Clagett introduced a bill in the U.S. House of Representatives for the establishment of 2 million acres around the upper Yellowstone as a "public park or pleasuring ground for the benefit and enjoyment of the people." Within two months the bill had passed both houses, and on March 1, 1872, President Grant signed it into law.[16]

Certainly Hayden's political prominence in Washington was important to creation of America's first national park. Another important factor was active support by the railroads, the most powerful lobby of the day. Part of the political success can be attributed to two artists who accompanied the Hayden expedition. One was wilderness photographer William Jackson. The other, whose participation was sponsored by Jay Cooke and the Northern Pacific Railroad, was painter Thomas Moran. During congressional consideration of the legislation to create the park, the photos of Jackson and the sketches of Moran (his epic oil paintings were not yet complete) were put on display in the Capitol rotunda. Moran described the Grand Canyon of the Yellowstone as "beyond the reach of human art." Modesty aside, the artistry of both Moran and Jackson helped to affix a visual image of Yellowstone's wonders in a way that written or oral testimony could not.[17]

Success was not merely a matter of enthusiastic park supporters. Equally important was the lack of concerted opposition. As events would soon show, the establishment of Yellowstone as the first national park did not represent a fundamental turn away from the prevailing robber-baron ethic of the Gilded Age. To many congressmen who voted for Yellow-

"Beyond the reach of human art." The art of Thomas Moran and the photography of William Jackson helped to convey to Congress the visual richness of the Yellowstone region.
Courtesy of the National Park Service, Yellowstone National Park.

William Jackson's photo of Old Faithful, the epitome of the "wonderful freaks of physical geography" that inspired creation of Yellowstone National Park. *Courtesy of the National Park Service, Yellowstone National Park.*

stone, the most compelling argument was not that the region's great beauty deserved to be protected but rather that the area was without economic value and therefore setting it aside *didn't matter*.[18] While a handful of farsighted advocates saw inherent worth in a natural reserve, it was benign indifference that carried the day.

Indifference, though, was no foundation for management. And from the vantage of the park itself, the effects would soon be shown to be anything but benign.

IF THE U.S. CONGRESS WAS LARGELY INDIFFERENT ABOUT THE NEWLY created park, part of the explanation lay in the Yellowstone region's extreme isolation—even in the era of a transcontinental railroad. It took George Bird Grinnell and the Ludlow-led soldiers more than two months to travel from Bismarck (the end of the slowly advancing Northern Pacific Railroad) to the park itself.

The steamboat *Josephine* carried the party 640 miles up the Missouri to the tiny outpost of Carroll, a town that Grinnell described as "two trading stores and half a dozen log cabins." Between Carroll and Yellowstone, wrote Grinnell, lay 300 miles of "a sort of debatable ground roamed over by half a dozen tribes of Indians all at war among themselves and most of them with the whites." On the day before Grinnell arrived in Carroll, Indian raiders had stolen every horse in town "except for one crippled animal that was unable to walk."[19]

Colonel Ludlow believed that his party was "not much in excess of one hundred miles" from the "hostile camps of Sitting Bull on the Yellowstone." The potential for violent contact with Indians—the Sioux in particular—resulted in tension throughout the trip. At one juncture Grinnell's party crossed paths with a detachment of soldiers who the day before had buried three of their comrades. The victims had been fishing in a creek when sixteen Sioux warriors swept out of a ravine and killed them. A large group of Crow, also camped in the area, took off in pursuit of the Sioux.[20]

The next day, as Grinnell watched, the Crow warriors returned from a successful attack on the Sioux, carrying scalps and leading the ponies of their dead rivals. One of the Crow had been killed, and Grinnell described the mourning ritual of the dead warrior's wife: "[T]he old woman dismounted from her horse, and walking over to a wagon that stood by the building, she drew her butcher knife, placed her little finger on the wagon wheel and by a quick stroke cut off the finger."

"There," said Charlie Reynolds to Grinnell. "You see that those people are really sorry for their loss."[21]

With the peril of Indian attack ever present during the long overland journey across Montana, Grinnell appreciated the steady leadership of

Reynolds. In his memoirs, Grinnell related an argument between Reynolds and another man over where to sleep for the night. They had just cooked supper over an open fire, and Reynolds wanted to move on, stop after dark, then sleep with no fire. When the other man objected, Reynolds said, "I don't mind getting killed, but I should hate to have somebody come along next year and kick my skull along the sand and say, 'I wonder what fool this was, who built a fire in this country and then slept by it.'"[22] Reynolds's perspective prevailed.

As he traversed the wild Montana of 1875, Grinnell focused intently on his mission as paleontologist and zoologist. In the official report he prepared at the conclusion of the expedition, Grinnell documented forty species of mammals that he "saw and identified either in life or by their remains." Buffalo—the core of the great northern herd—were abundant through most of his journey: along the banks of the Missouri, on the overland passage through Judith Gap, and in Yellowstone Park itself. Where buffalo were absent, signs of their recent presence were everywhere. Buffalo trails crisscrossed the prairie floor, and in places, according to Grinnell, some paths had been trodden to a depth of two feet.[23]

Colonel Ludlow also discussed the buffalo in his official report. As the expedition churned up the Missouri, Ludlow described how a herd of seventy-five or eighty animals plunged into the river in front of their steamer. The steamer intercepted them, and the animals turned back only after the leaders "struck the boat with their heads." Unlike generations of plinkers, though, Ludlow and his men resisted the temptation to shoot: "It would have been butchery to kill them, especially as we did not need the beef, and they were allowed to escape unhurt."[24]

On August 9, 1875, the Ludlow expedition reached Fort Ellis, a small outpost near present-day Bozeman. They spent a day preparing supplies and equipment for their foray into the park before proceeding again on August 11, traversing into the stunning valley of the Yellowstone River (known today, aptly, as Paradise Valley). For two days the party hugged the river, its pace casual enough to allow admiration of the trout and grayling that "competed successfully for the fly." On August 13, near Mammoth Hot Springs, the Ludlow party entered Yellowstone National Park.[25]

In their official reports, both Ludlow and Grinnell would echo some of the experiences of earlier explorers: They marveled at the grand panoramas, the geological oddities, the wildlife. But for both men, the journey also exposed them to something jarringly unexpected: The hide hunters, too, had discovered the upper Yellowstone. "It is estimated that during the winter of 1874–1875," wrote Grinnell, "not less than 3,000 elk were killed for their hides alone in the valley of the Yellowstone. . . . Buffalo and mule-deer suffer even more severely than the elk, and the antelope nearly as much."[26]

As an introduction to his report for the Yellowstone expedition, Grinnell drafted a single page he titled "Letter of Transmittal," which included the following:

> It may not be out of place here, to call your attention to the
> terrible destruction of large game, for the hides alone, which is
> constantly going on in those portions of Montana and Wyoming
> through which we passed. Buffalo, elk, mule-deer, and antelope
> are being slaughtered by thousands each year, without regard
> to age or sex, and at all seasons. Of the vast majority of the
> animals killed, the hide only is taken. Females of all these
> species are as eagerly pursued in the spring, when just about to
> bring forth their young, as at any other time.[27]

Colonel Ludlow too would document the early abuses of Yellowstone National Park. A year earlier in *Report of the Reconnaissance of the Black Hills*, the colonel had lodged a prescient warning against taking the Black Hills from the Sioux. ("The Indians have no country farther west to which they can migrate.") Now, in his report on Yellowstone, Ludlow aimed his outrage against the small army of vandals who, by 1875, already had made their mark on the park. "The only blemishes on this artistic handiwork had been occasioned by the rude hand of man."[28]

Indeed, when the Ludlow expedition reached Old Faithful, they found a large party already there, including a "whiskey-trader snugly ensconced beneath his 'paulin, spread in the shelter of a thick pine." More outrageous still was what Ludlow described as a swarm of tourists who "prowled about with shovel and ax, chopping and hacking and prying up great pieces of the most ornamental work they could find; women and men alike joining in the barbarous pastime."[29]

GRINNELL'S AND LUDLOW'S CONCERNS FOR WILDLIFE AND OTHER natural resources surely were far ahead of their time, but they did not spring from whole cloth. For more than a century, in the United States and in Europe, diverse strains of thought had been weaving a context against which the notion of protecting the wild could begin to take hold.[30]

Euro-American appreciation for the natural world began with scientific advances of the sixteenth and seventeenth centuries, including a growing understanding of the solar system, gravity, and oceans. In religion, movements such as the "deism" of the seventeenth and eighteenth centuries posited a close association between God and nature. In philosophy, the Romantic movement of the eighteenth and nineteenth centuries celebrated the beauty, the mystery, and the very wildness of the natural world.

Writers and artists would play a significant role in shaping how Americans thought about the natural world. Ever so slowly, the traditional, conquest-oriented conceptions of the relationship between mankind and nature began to change. In the 1820s and 1830s, for example, James Fenimore Cooper wrote his *Leatherstocking Tales*, a series including *The Deerslayer* and *The Last of the Mohicans*. Cooper's hero, Natty Bumppo—better known by his Indian name of Hawkeye—was a man wholly attuned to the wild world around him, a world that embodied the Romantic vision of nature as a source of beauty, character, and even the sacred. To waste resources, in Hawkeye's world, was an act of villainy: "They scourge the very 'arth with their axes. Such hills and hunting grounds as I have seen stripped of the gifts of the Lord; without remorse or shame!"[31]

Cooper wrote a lengthy death scene of his most famous character in *The Prairie*. As an old man, Hawkeye has retreated from his beloved, beleaguered forests to the last frontier—the wide prairie west of the Mississippi. As he looks back on his long life, Hawkeye aches for the wilderness of his youth:

> *What this world of America is coming to, and where the*
> *machinations and inventions of its people are to have an end,*

the Lord, he only knows. I have seen in my day, the chief, who
in his time, had beheld the first Christian that plac'd his wicked
foot in the regions of York! How much has the beauty of the
wilderness been deformed in two short lives! . . . And where am
I now! . . . towns, and villages, farms and high ways, churches,
and schools, in short all the inventions and deviltries of man are
spread across the region![32]

Hawkeye's perspective stood in stark contrast to the belief of most nineteenth-century Americans in the necessity—even the sanctity—of mankind's conquest of the wild.

Visual artists of the early nineteenth century were also portraying the world in new ways. Central to their vision was the beauty of nature. American Thomas Cole, whose work would inspire the Hudson River School, reduced the humans in his epic landscape paintings to minuscule scale—or left them out altogether. Here was a view of the world in which mankind was but a small part of a far broader canvas.[33]

Not surprisingly, Romantic artists searching for inspirational scenes of natural beauty soon turned to the American West. Indeed it was Romantic painters who gave Americans their first views of the West. Painting in the 1830s, George Catlin and Karl Bodmer were among the earliest artists to stir the American imagination about horizons west of the Mississippi, often portraying Romantic visions of the Indians and their "noble savage" lifestyle. John James Audubon would focus his talent and attention on birds and other wildlife, celebrating the natural world with an appreciation that put him comfortably in the Romantic movement.

Landscape artists would also find inspiration in the West. Thomas Moran, beginning with his participation in the 1871 Hayden expedition of Yellowstone, would spend a decade painting the grand vistas of the Rocky Mountains. Moran's work was significant not only because it helped to convince Congress to create Yellowstone National Park but more broadly because Moran and other Romantic painters artists helped a generation of Americans to see in nature something inherently beautiful.[34]

The growing American appreciation for nature's beauty began, toward the middle decades of the nineteenth century, to take on another important attribute. In his important work *Wilderness and the American*

Mind, historian Roderick Nash wrote that while *conquest* of wild country would continue to be a source of American pride, "by the middle
decades of the nineteenth century wilderness was recognized as a cul-
tural and moral resource and a basis for national pride."[35]

America in the first century of its existence cast a frequent and inse-
cure eye toward Europe. With Europe's millennia of civilization and
culture, how could the upstart United States hope to measure up? The
answer, for a growing number of American intellectuals, was to empha-
size what the United States had that Europe did not—wild places. As
the nineteenth century progressed, more and more Americans came to
realize that wilderness contributed not only to their nation's treasure but
also to its character. Wilderness defined America and made it different
from Europe. Indeed it was wilderness—and a frontier spirit—that
defined the American people themselves, setting them apart from their
ancestors of the Old World.[36]

THERE IS AN ATTRACTIVE NOTION—ATTRACTIVE BUT FALSE—THAT
the creation of Yellowstone National Park in 1872 was the capstone of
a century's worth of percolating environmental thought. It is true that
developments in religion, science, philosophy, literature, and art all contributed to the creation of a growing popular appreciation for the natural world. It is also true that these same developments helped to create a
vanguard—soon to include George Bird Grinnell—of advocates for
wild places and wild things.

What had not yet happened in 1875, however, was the translation of
this favorable sentiment toward nature into political power.

In the creation of Yellowstone, the vanguard was able to push through
establishment of the first national park not because of a broad-based desire
to protect wild places—but rather because the indifferent majority in
Congress was convinced that the area could be set aside without economic sacrifice. Even for those who advocated the park as a necessity to
protect nature, the focus was not on the region's mountains, forests, rivers,
lakes, and wildlife. Rather they viewed Yellowstone as a repository of curiosities, an outdoor museum of "wonderful freaks of physical geography."
For most, even ardent supporters, the park's significance resided in its bub-
bling mud pots, gigantic falls, and spouting geysers.[37]

Certainly the establishment of the nation's first national park represented a significant historical event, but it was not the beginning of a new era nor the triumph of an earlier one. Setting aside the remote land for Yellowstone was the easy part; it didn't even require an appropriation. Truer tests of the nation's underlying sentiment toward the natural world would arise when protection came at some price—be it federal expenditure or lost economic opportunity. Those tests would prove Congress's support for nature to be superficial at best.

Indeed, congressional indifference was a primary reason that the Ludlow expedition found tourists at Old Faithful hacking up geyserite: No measures had been taken to prevent vandalism. Three years after the establishment of the park, not a single U.S. official even resided within its boundaries. Congress's budgetary priorities are telling. In 1874, the year before Grinnell explored Yellowstone, Congress appropriated $10,000 to commission a Thomas Moran painting of the Grand Canyon for display in the Senate lobby. In the same year, though, a request for $25,000 to protect park resources died before reaching the floor of either house. In fact, in the first six years of the new national park's history, Congress would refuse to spend a single penny on Yellowstone's maintenance or protection. The message seems clear: The effect of a century of growing environmental awareness was a nation prepared to appreciate artistic representations of natural beauty but unwilling to pay a price to protect the real thing.[38]

THE PROFOUND EFFECT OF THE YELLOWSTONE EXPEDITION ON LUDLOW and Grinnell is apparent in their official reports in the year after their journey. Ludlow, in language that sounds more like a Romantic novelist than a government bureaucrat, described Yellowstone as a place "where the hidden pulses can, as it were, be seen and felt to beat, and the closely-written geological pages constitute a book which, being interpreted, will expose many of the mysterious operations of nature."

Ludlow expressed his hope for the park's protection "from the vandalism from which it has already suffered." In a strikingly controversial recommendation (a decade ahead of its time), Ludlow also called for the removal of Yellowstone from the jurisdiction of the Department of the Interior in order to place it under the protection of the Department of War.[39]

Grinnell, whose report focused on wildlife, would conclude his "Letter of Transmittal" with an explicit warning: "It is certain that, unless in some way the destruction of these animals can be checked, the large game still so abundant in some localities will ere long be exterminated." Grinnell's letter would be published by the Government Printing Office in 1876 as part of the U.S. Army report on the Ludlow reconnaissance, thus becoming one of the first official protests of its kind.

On his first trip west in 1870, George Bird Grinnell had discovered the passion for wild places that would animate the rest of his life. In the Yellowstone expedition of 1875, Grinnell now married his passion with purpose. But to prevent the extermination he foresaw was a far more ambitious task than any previously undertaken in support of the natural world. Saving the buffalo would require something the establishment of Yellowstone had not—raw political power.

To succeed, Grinnell would need to create something that did not exist in the 1870s—a national political constituency for wildlife and wild places. And not just any constituency. The goal of protecting the buffalo, in a few short years, would come into conflict with the vested interests of railroads, mining, and real-estate developers, not to mention the skin hunters and the broader industry they supported. In short, to save the buffalo, Grinnell would need a political force that could stand toe-to-toe with the most potent powers of the Gilded Age.

CHAPTER EIGHT

"A Weekly Journal"

Forest and Stream, *a weekly journal: Devoted to Field and Aquatic Sports, Practical Natural History, Fish Culture, the Protection of Game, Preservation of Forests and the Inculcation of Men and Women of a Healthy Interest in Out-Door Recreation and Study.*

—MASTHEAD,
Forest and Stream

I n the winter of 1877, J. Wright Mooar and his partners harvested 3,700 buffalo hides in the Texas Panhandle, transporting them by wagon to market in Fort Worth. It was a good showing, but as Mooar knew, change was coming fast for buffalo runners of the Southern Plains.

Earlier that same year, the Mooar outfit had been forced to move north after receiving a report that "a large herd of cattle was coming into the country." By the spring of 1878, Mooar could see what the future held. He continued to hunt buffalo—when he could find them— but he also began to buy land and cattle. By 1879, Mooar remembered, "the hunters were leaving for the mining states, or seeking other lines of business, as they realized the great hunting days were over." Mooar took one last load of cured buffalo meat to the mining camps of Arizona. After selling it, he returned to Texas and took up ranching full time.[1]

By 1879 the southern herd was gone, millions of buffalo decimated in the eight bloody years between 1871 and 1879. The surviving buffalo on the North American continent now comprised two groups—the northern herd of the United States on the broad Montana–Dakota grasslands and a Canadian herd in Alberta and Saskatchewan.

The buffalo, at least, had managed to outlive General Custer. On June 25, 1876, as he hunted for the Indians who had refused to report to reservations—Custer found them. In southeastern Montana, five tribes of Sioux and one of Cheyenne had converged in an enormous encampment along the meandering Little Bighorn River. An estimated 7,000 Indians inhabited the camp, an unprecedented number that included 1,800 fighting men. Leading 210 soldiers of his vaunted Seventh Cavalry, Custer charged into the valley. An hour later all of them were dead.

One of the men who died that day was a civilian scout, Charlie Reynolds, cut down as he attempted to cover the retreat of Major Marcus Reno. One man who had almost been there was George Bird Grinnell.[2]

Back in New Haven after returning from the Yellowstone expedition, Grinnell had received a telegram from Custer in May of 1876. The general invited Grinnell "to be his guest on an expedition to be made in conjunction with other troops up the Yellowstone and toward the Big Horn Mountains." Grinnell, as he remembered, was extremely busy with his Peabody Museum responsibilities and other work for his mentor, Professor Marsh. Given the nature of Custer's 1876 mission, Marsh doubted that there would be adequate time for scientific exploration. Grinnell telegraphed Custer with his regrets. It was the first time he had ever declined an opportunity to go west.

"Had I gone with Custer I should in all probability have been mixed up in the Custer battle," wrote Grinnell, "for I should have been either with Custer's command, or with that of Reno, and would have been right on the ground when the Seventh Cavalry was wiped out."

In addition to Grinnell's study for his doctorate and his work at the Peabody Museum, the 26-year-old had assumed another responsibility in 1876. Three years earlier, amid the turmoil of his family's financial crisis, Grinnell had begun an informal association with an influential new sportsmen's journal. "The winter was one of great suffering for the family," he later remembered, "and in my effort to divert my mind from the misfortune that had happened, I began to write short hunting stories for *Forest and Stream*."[3]

In the spring of 1876, probably just prior to receiving his invitation to accompany Custer, *Forest and Stream* made Grinnell a tempting offer. The journal needed a new editor for its natural history page. Grinnell, who combined a passion for hunting and the outdoors with a Yale pedigree, was an obvious choice. He accepted the job and so began what would be a 35-year tenure with the magazine.

For $10 a week, Grinnell's responsibility was "to furnish a page or more of material weekly, to write reviews of natural history books, and generally to make myself efficient." He worked from New Haven, sending his material to New York by mail while continuing his doctoral studies with Marsh.

His audience of hunters and anglers would rise as the first political constituency in support of the wild.

FOR NON-SPORTSMEN IN THE EARLY TWENTY-FIRST CENTURY, THE notion of such groups as guardians of the environment may seem counterintuitive. History's record, though, is unambiguous. Decades before the rise of such important groups as the Sierra Club, decades before Gifford Pinchot's "scientific forestry," decades before the presidency of Theodore Roosevelt—sportsmen were virtually the only organized group to support the wild.[4]

Anglers perhaps deserve the designation as the first conservationists. (The term *conservationist* would not actually enter the popular lexicon until Roosevelt's presidency.) In 1734, four decades before the Declaration of Independence, a New York City statute attempted to limit fishing in freshwater ponds to "angling with angle-rod, hook, and line." The law's aim was to curtail the devastating practice of netting. New York, because of its population, would be one of the first places in the nation to experience many of the conflicts between mankind and nature.[5]

Massachusetts, another population center, also passed early laws in an effort to protect fisheries. Daniel Webster, the future titan of the U.S. Senate (and avid fisherman), spent a short term in the Massachusetts House of Representatives. While there, his primary achievement was the passage of an 1822 law providing that "no man shall catch trout in any other manner than with the ordinary hook and line."[6]

Webster's statute would fail for the same reason that most early efforts to protect wildlife failed, and for the same reason that tourists and market hunters in 1875 were plundering the new Yellowstone National Park: Conservation laws, once passed, were rarely enforced. In 1833, a Boston physician-angler would complain that Webster's law was "disregarded," and that "Factories and saw-mills have done their part towards the work of [the trout's] extermination, and the destructive *net* bids fair to the rest."[7]

New York City's efforts to protect its fish would also fail. In 1865, Robert Barnwell Roosevelt—Theodore Roosevelt's uncle—wrote that "streams in the neighborhood of New York that formerly were alive with trout are now totally deserted. The Bronx, famous alike for its historical associations and its once excellent fishing, does not now seem to hold a solitary trout or indeed fish of any kind." A century before Norman McClain's *A River Runs Through It*, the elder Roosevelt would also castigate the fisherman who dared chase a trout with a worm instead of "the artificial fly." Such a man, according to Roosevelt, acted "at great risk to his reputation."[8]

The fishermen's most significant contribution to the early development of a conservation ethic came in the person of a bespectacled attorney named George Perkins Marsh (no relation, apparently, to Professor Othniel Marsh), an avid angler during his boyhood in Vermont. In the 1840s, Marsh had been one of the legislators influential in the establishment of the Smithsonian Institution. After spending several years abroad in ambassadorial positions, Marsh returned to Vermont in 1853. There, he accepted an appointment as railroad commissioner, statehouse commissioner, and fish commissioner.[9]

Vermont, in Marsh's lifetime, had seen the extinction or near extinction of several species of fish against a backdrop of deforestation and the extensive construction of dams. Concerned fishermen, hoping to restore their beloved sport, convinced the state legislature to "enquire into the present state of the discoveries which have been made in relation to the artificial propagation of fish." The governor tasked Fish Commissioner Marsh with investigating.[10]

Marsh's report laid blame for diminished fish numbers at the feet of both anglers and industry. As for anglers, he decried the taking of fish "at the spawning season, or in greater numbers at other times than the natural increase can supply." As for industry, Marsh described how "the erection of sawmills, factories and other industrial establishments on all our considerable streams, has tended to destroy or drive away fish."

Even more forward-looking was Marsh's concern for what he described as the "more obscure causes" impacting Vermont's fisheries. In particular, Marsh focused on the effect of clearing forests for agriculture. "Many brooks and rivulets, which once flowed with a clear, gentle, and equable stream through the year, are now dry or nearly so in the summer, but turbid with mud and swollen to the size of a river after heavy rains or sudden thaws." Marsh believed that such conditions had a major impact on the ability of fish to reproduce. Though he did not use the word *habitat*, Marsh described his concern for the impact on the "gravelly reach" that fish require "as an appropriate place for the deposit of eggs."[11]

Marsh's fish report focused more on a description of the problem than on suggestions for solutions. But his work underscored the role of sportsmen at the leading edge of efforts to understand the relationship between humans and their environment. Even more significantly, Marsh's research on the Vermont fisheries would lead directly to the trailblazing work for which he is best known today.[12]

In 1864 Marsh published *Man and Nature*, subtitled *Physical Geography as Modified by Human Action*. One historian points to the book as no less than the "fountainhead of the conservation movement." Wallace

George Perkins Marsh: Wallace
Stegner called Marsh's *Man and
Nature* the "rudest kick in the face
that American initiative, optimism
and carelessness had yet received."
*Courtesy of the National Park Service,
Yellowstone National Park.*

Stegner succinctly placed Marsh's work in the context of its times, calling *Man and Nature* the "rudest kick in the face that American initiative, optimism and carelessness had yet received."[13]

Read from a distance of a century and a half, *Man and Nature* remains a stunningly contemporary work. Marsh opened his book with a provocative examination of history, concluding that the fall of many ancient civilizations could be traced, in significant part, to the "ignorant disregard of the laws of nature."[14]

As most of the nation celebrated the conquest of the wild, Marsh warned that "[m]an has too long forgotten that the earth was given to him for usufruct alone, not for consumption, still less for profligate waste." As one example of the "terrible destructiveness of man," Marsh pointed to the "chase of large mammalia and birds for single products," mentioning specifically the slaughter of the buffalo "for his skin or his tongue."[15]

Marsh also looked to the future. For areas already harmed by people, Marsh was the first to call for what has become a cornerstone of modern environmentalism—restoration. Marsh called it "restoration of disturbed harmonies," and what was needed, Marsh argued, was a new kind of pioneer. "In reclaiming and reoccupying lands laid waste by human improvidence or malice . . . the task of the pioneer settler is of a very different character. He is to become a co-worker with nature in the reconstruction of the damaged fabric which the negligence or the wantonness of former lodgers has rendered untenantable."[16]

Marsh's *Man and Nature* was first published when George Bird Grinnell was a boy of 14. Though Grinnell's writings make no specific reference to the book, it seems certain that he was familiar with Marsh's work. *Man and Nature* would have a recognized impact on policy as early as 1873, when it catalyzed the creation of the United States' first national forestry commission.[17] Certainly Marsh, with his close ties to the world of fishing, helped to establish the intellectual foundation for many positions that Grinnell would later help to popularize.

Like anglers, hunters could also boast a long connection with the principles that would evolve into full-fledged conservationism. The concerns of hunters, arising first in the populated East and South, stemmed from their personal experiences of arriving at favored haunts to discover that once-plentiful game was now gone.[18]

One scholarly analysis looked at two prominent antebellum sporting journals, *Spirit of the Times* and *American Turf Register*. The primary focus of both journals was the popular sport of horse racing. Hunting and fishing, though, received significant coverage, including repeated advocacy for strong game laws, virulent antipathy toward the excesses of market hunters, sophisticated discussion of ideas for fish and wildlife propagation, and "general expressions of concern" about the destruction of the wild places on which wildlife depended.[19]

The opinions expressed by the readers of these journals are quite surprising. One Virginia subscriber, for example, had the temerity to challenge no less a sacred cow than tobacco farming. "I must premise that in this section of the state, where much of the native forest has been cut down, and the lands cleared, in consequence of their being particularly adapted to that curse of the Ancient Dominion, *tobacco*, the deer have become very scarce."[20]

Rather than simply pine for the good old days, hunters began to emerge as an activist force for change. Early journals focused repeatedly on appeals to their readers' "gentility and sense of honor to rectify certain abuses." One column in an 1830 edition of *American Turf Register* framed the issue in almost biblical terms: "We think it behooves every real sportsman to refrain from doing any act which he would wish to be secret; and although there may be some speciousness in the excuse, that 'if I don't kill the birds now, others will,' still two wrongs can never make a right."[21]

Unfortunately, as ethical hunters came to realize, such personal honor was not sufficiently widespread to prevent abuses. By the 1840s, sportsmen, often organizing into local clubs, were beginning to flex their political muscle. In state and local governments, hunters pushed for and sometimes achieved the passage of progressive game laws. In Pennsylvania, for example, an 1858 law limited the hunting of small game to the fall. In 1859, New York sportsmen called a convention "for the purpose of discussing and devising means for united action throughout the State for a revision of the present senseless and inefficient Game and Fish Laws." In the same year, Ohio sportsmen pushed through a statute that limited the hunting of game species to the fall and also made illegal the common practice of trapping (versus shooting) birds. [22]

Though hunters succeeded in putting a significant number of game protection laws on the books, in few cases did they solve the vexing issue of enforcement. Indeed enforcement was one of the issues where Grinnell would ultimately make the greatest contribution.

In *Man and Nature*, George Perkins Marsh described the political barrier that often impeded enforcement of game laws in the United States: a uniquely American antipathy toward the regulation of hunting. Americans in the nineteenth century drew a strong association between game laws and what Marsh called the "abuses of feudalism." Under the traditional European systems, game laws were a mechanism to protect hunting grounds for a rich or noble elite. (Think Sherwood Forest.) "[T]he evils the peasantry had suffered from the legislation which protected both [the forest] and the game it sheltered, blinded them to the still greater physical mischiefs which its destruction was to entail upon them." For the poor hunter who sought to feed himself or his family by poaching on these European reserves, punishments could include the death penalty.[23]

Against this feudal backdrop, proposals for game laws in the United States were often perceived as an effort to impose an unacceptable brand of European elitism on the American continent. Before the Civil War, advocates of game protection laws had not yet discovered the political formula that would allow them to advocate for the protection of wildlife without alienating a large group who saw unrestricted hunting on public land as a fundamental right of American democracy.[24]

AMERICA'S PREOCCUPATION WITH THE CIVIL WAR WOULD RESULT IN a significant pause in the development of conservationist thinking. (Marsh's *Man and Nature* was a notable exception.) For many sportsmen's journals, including *Spirit of the Times*, the war was a death knell. One impact of the conflict was a severance of economic relations between North and South—including postal service—with obvious effects on the magazine industry. Within the South itself, the magazine industry virtually evaporated as a result of high postage rates and the shortage of such vital materials as ink.[25]

The end of the Civil War and the economic boom of the early Gilded Age resulted in a "mania of magazine starting." Having endured the war, the nation was ready for all manner of diversion, and magazines in the second half of the nineteenth century were both popular and influential. In 1865, the United States could count approximately 700 periodicals (not including newspapers). By 1885 there would be more than 3,000.[26]

The diversity of titles was remarkable, from the *Alaskan Herald* to *Zion's Watchman* and everything in between. Magazines organized around region, political and religious affiliation, ethnicity, occupations, the sciences, literature and art, and by every pastime and hobby imaginable.

One of the most popular magazine genres, picking up where the handful of influential prewar magazines left off, was sports and outdoor life. Some periodicals focused on such antebellum sports as horse racing (for example, *Spirit of the Times*, back from the dead) and boxing (*National Police Gazette*). Other new titles focused on postbellum fads such as baseball (*Ball Players' Chronicle*) and biking (*Wheelman*).

Among the sporting journals, no topics received more attention than hunting and fishing, with more than twenty titles devoted to them. In part because of the influence of the Romantic movement, more Americans were taking to the outdoors for recreation. "America's play-house is all-out-of-doors," proclaimed the *Chicago Field*, one popular sportsmen's journal, "and out-door sports are growing more popular."[27] Of the score of such journals, a handful ultimately would distinguish themselves through their continued focus on the conservation of wildlife: *American Sportsman* was the first, established in 1871; two, *Forest and Stream* and *Field & Stream*, were established in 1873; *American Angler* followed in 1881.

Together, these journals gave hunters and anglers a mechanism for communicating with one another, a sort of weekly rallying point. The result is the fostering of what historian John F. Reiger has identified as a "rapid growth of group identity." And in the rough-and-tumble political battles that would unfold over the next two decades, no single resource would prove more effective than Grinnell's *Forest and Stream*.[28]

FOREST AND STREAM WAS THREE YEARS OLD WHEN GRINNELL, IN 1876, began editing the natural history page. The magazine's founder was Charles Hallock, a blue-blooded New Yorker who envisioned his publication as far more than a collection of anecdotes about hunting and fishing—though it included plenty of those. The journal's masthead threw down its aspirations like a manifesto: "*Forest and Stream*, a weekly journal: Devoted to Field and Aquatic Sports, Practical Natural History, Fish Culture, the Protection of Game, Preservation of Forests and the Inculcation of Men and Women of a Healthy Interest in Out-Door Recreation and Study." Here, explicitly declared, was a publication whose raison d'être included the protection of wild animals and wild places.

In *Forest and Stream*'s first issue, Hallock explained the type of reader at which he aimed. The mere act of hunting or fishing did not qualify a person as a true sportsman. "It is not sufficient that a man should be able to knock over his birds dexterously right and left, or cast an inimitable fly." To be a true sportsman, according to Hallock, required a familiarity with "the living intelligences that people the woods and the fountains." As a model, Hallock pointed (without irony) to the "savages"—who "familiarized themselves with the habit of every form of animal life. Under each decaying leaf, in each blade of grass or rolling log, they discovered a microcosm." Foreshadowing the importance of Grinnell's column on natural history to the mission of the magazine, Hallock wrote that "[a] practical knowledge of natural history must of necessity underlie all attainments which combine to make a thorough sportsman."

For the true sportsmen whom Hallock hoped to attract, he intended for *Forest and Stream* to serve as the "recognized medium of communication." Hallock's inaugural issue would also make clear that the establishment of this virtual community had a political purpose. Reflecting the mission proclaimed in the masthead, Hallock promised his readers that "[f]or the preservation of our rapidly diminishing forests, we shall continually do battle." *Forest and Stream* would "occupy an independent position, and throw our efforts in behalf of a competent reform. We shall perhaps even *clamor* for it."[29]

Hallock wasted no time in demonstrating *Forest and Stream*'s pugnacious tone. Not only would the magazine advocate for the ideals of the true sportsman but it would also deride those who ignored them—a magazine version of the stocks in the village square. In one early issue, for example, Hallock flayed the Reverend W.H.H. Murray of Boston under an editorial headlined "Killing Game Out of Season." The good reverend had visited the forests of the Adirondacks in the summer of 1867, hunting and fishing in violation of closed-season laws. Demon-

NEW YORK, THURSDAY, AUGUST 14, 1873.

FOREST AND STREAM.

On the fair face of Nature lit as smar,
And drawn by lapsing streams and drooping wood ;
'Twixt the dark furrow where immortal roots
Since the remotest days have cast their shade ;
Ponder by flowing stream and ocean tides,
And mark the varied forms of life they hold,
Mark the wild game so dear to hunter's heart,
The swarming fins that skim the salty deeps,
The birds that haunt the woodlands and the plains,
The fish that swim the seas, the lakes, the streams
And tempt the thoughtful angler to their snug ;

...

Isaac McLellan.

ANTICOSTI.

THE JOURNAL OF A NAVAL OFFICER.

Till within the last few years, the island whose name stands at the head of this article, has been to the great majority, what may well be termed "*terra incognita.*" Heard of but seldom, and then only in connection with disaster, it is perhaps no wonder that the island has from the earliest times acquired an ill-omened reputation, from the long list of ships whose timbers have found a last resting place on its shores, and whose names swell the ghastly record of missing vessels. Situated at the extreme of the Gulf of St. Lawrence, and lying in the centre of the highway of that tide of shipping which sets in towards the Labrador coast and the navigable season, the heavy storms which sweep the Labrador coast, and dense fogs which then prevail in those latitudes, coupled with the insidious currents, should fare many a ship within the line of reefs which circle its shores. The mind is however apt to magnify dangers it cannot fully comprehend, and there is little reason to doubt that were the island better known, many of the wild reports, amounting almost to superstition, with which fancy is wont to invest it, would be dispelled.

...

It was not till the year 1829 that the government of Canada was roused into activity by the general consternation which was displayed when the news of the loss of the "Granicus," with all hands, and the almost incredible horrors undergone by those who had escaped shipwreck only to meet a more horrible death, became known. Towards the end of 1828, a ship began to be felt for the safety of the "Granicus," a large ship on her passage to Quebec, and as the winter months rolled on, and no tidings reached those who anxiously awaited some loved one's return, she was probably put down in that long list of ships which have foundered in mid ocean, and whose fate must for ever remain a mystery.

...

strating the broader lack of concern about enforcement, the Reverend Murray proceeded to publish a book in which he bragged of his exploits: "I have deceived the finny tribe and killed and eaten a half score of deer slain by my own hand."

"Among all true sportsmen," wrote Hallock in response, "there is a bond of sympathy, one touch of which makes the fraternity akin, and within this charmed circle, Mr. Murray has not yet been admitted, and never will be, so long as he continues to slaughter game out of season." Hallock contrasted Murray's ethic with that of James Fenimore Cooper's Hawkeye: "I kill no bird unless it has a chance for its life on the wing; and no four-footed game except in its season of health, and possessed of all the advantages which God has given it for escape."

"But take heart," continued Hallock to the Reverend Murray, "you are only one of the many who have helped within a few years to almost annihilate the game of the forest and streams of the northern wilderness." Finally came a promise: "Against this waste we shall wage a constant war."[30]

In 1873, as the jarring impact of commercial buffalo hunting on the Southern Plains was just beginning to be reverberate, *Forest and Stream* became one of the first publications to publicize the slaughter—and to protest against it. In an October editorial titled "Destruction of the Buffalo," Hallock detailed the "wanton waste of animal life and food in the Far West, especially of the buffalo, which are slaughtered for their skins by the hundred thousand, and left to rot in their tracks."

Hallock noted a Colorado law that attempted to stop the waste, requiring that no person should "leave any edible portion of such game so killed to waste, but shall take care of and preserve or bring into market each and all parts of such game that are edible." Recognizing that enforcement would be essential, Hallock called for the employment of "detectives" to travel out into the hunting grounds. Later editorials would urge the adoption of a closed season during which buffalo could not be hunted. Presaging modern management of migratory animals,

The inaugural issue of *Forest and Stream* from August 14, 1873.
Courtesy of Archives and Special Collections, University of Montana-Missoula.

Hallock also urged governors in the West to cooperate in enacting solutions that crossed the artificial barriers of state lines.[31]

In December 1873, *Forest and Stream* published the account by George Bird Grinnell of his buffalo hunt in Kansas with the Pawnee. Regarding the Indians' use of the meat, Grinnell described how every ounce would be saved. "How different would have been the course of a party of white hunters had they had the same opportunity," wrote Grinnell. "They would have killed as many animals, but would have left all but enough for one day's use to be devoured by the wolves or to rot upon the prairie."[32] Hallock would later highlight Grinnell's description of the Pawnee hunt as a pointed contrast to the commercial slaughter:

> *If the savage has an inordinate appetite, he only kills to eat and not for the amusement of slaughter. His instinct teaches him that did he destroy indiscriminately the animal life of the prairie, he would deprive himself in time of his only food. The Indian kills only what he wants for immediate consumption and in some cases prepares his buffalo meat, by drying it, for future use.*[33]

Even more surprising than *Forest and Stream*'s early attack on commercial hunting is the magazine's frontal assault on the strategy of killing the buffalo as a means of subjugating the Indians. In April 1874, as Custer prepared for his fateful exploration of the Black Hills, *Forest and Stream* editorialized that the "solution of the Indian question by 'killing all the buffalo,' may as well be cast aside as nonsense." Hallock instead expressed the hope that "before many years, a just, honest and efficient policy may be pursued towards the Indian, and that we can conscientiously aid in the increase of the buffalo instead of furthering its foolish and reckless slaughter."[34] An admirable aspiration, but one that would soon be proven naïve.

IN THE FOUR YEARS DURING WHICH HE SERVED AS EDITOR OF *Forest and Stream*'s natural history page, George Bird Grinnell continued his work with Professor Marsh at the Peabody Museum and still somehow managed to plug away on his doctorate. Grinnell also continued his regular, extended trips to the West.

As they had in the past, Grinnell's adventures out West continued to provide rich fodder for his writing. In the spring of 1877, he traveled to northwestern Nebraska to hunt antelope and elk with his friend Lute North. Lute's brother Frank had gone into the cattle business with Bill Cody, and Grinnell stayed as a working guest at their ranch. The ranch, according to Grinnell, consisted of "a couple of tents stuck up on the edge of the alkaline lake which was the head of the Dismal River." In 1878, Grinnell and his brother hunted fossils in the Como Bluffs region of Wyoming. The next year, he hunted in the North Park of Colorado (today part of Rocky Mountain National Park).[35] On one of these trips west (his memoirs are unclear) Grinnell caught a bad case of Rocky Mountain spotted fever. The tick-borne disease would nearly cost him his life, and his recovery required seven weeks in bed.[36]

In addition to the adventure he craved, Grinnell's decade of travel to the West had connected him in a profoundly personal way to a critical period in history. On his first trip west, in 1870, Grinnell's train had twice been halted by herds of buffalo. There were no longer buffalo in the Colorado of 1879, but Grinnell did find herds of cattle, some of them behind fences of barbed wire.[37]

Grinnell described his 1879 journey for the readers of *Forest and Stream* with a degree of nostalgia that would normally seem surprising for a man only 30 years old. "The rapidity with which our western country is settling up," he wrote, "impresses me more and more each year."[38] The breadth of change he had witnessed in the span of only ten years was breathtaking.

By 1879, the Indian wars on the Northern Plains were effectively over. The victory of Sitting Bull and Crazy Horse in the battle on the Little Bighorn might have been Custer's last stand, but it was the Indians who soon lost the war. Indeed by 1879 Crazy Horse had been dead for two years, murdered after surrendering himself and his people at the Red Cloud Agency. Chief Joseph of the Nez Perce had failed in his effort to lead his people to freedom, turning over his rifle with the haunting promise to "fight no more forever." Sitting Bull and a small band, the last Sioux holdouts, had taken temporary refuge in Canada. Driven by starvation, Sitting Bull and his followers would finally submit to the U.S. Army in 1881.

Grinnell described the pattern that followed the end of hostilities with the Indians: "[A]s soon as any section becomes safe, the Indians having been driven off, the cattlemen begin to drive their herds into it, and before long, one hears complaints that there are too many cattle there." Heavy industry had also come to the Rocky Mountain West. By 1879, the towering smokestacks of large-scale silver-mining operations were spewing arsenic-laden smoke into the skies over places such as Leadville, Colorado; Virginia City, Nevada; Butte, Montana; and Park City, Utah.

With all that he had witnessed, with all that he had experienced—it should not be surprising that Grinnell, in describing his 1879 trip to the readers of *Forest and Stream*, would write with palpable dread:

> *As I look back on the past ten years and see what changes have taken place in these glorious mountains since I first knew them, I can form some idea of the transformations which time to come will work in the appearance of the country, its fauna, and its flora. The enormous mineral wealth contained in the rock-ribbed hills will be every year more fully developed. Fire, air, and water working upon earth, will reveal more and more of the precious metals . . . Towns will spring up and flourish, and the pure, thin air of the mountains will be blackened and polluted by the smoke vomited from the chimneys of a thousand smelting furnaces; the game, once so plentiful, will have disappeared with the Indian; railroads will climb the steep sides of the mountains and wind through their narrow passes, carrying huge loads of provisions to the mining towns, and the returning trains will be freighted with ore just dug from the bowels of the earth; the valleys will be filled with fattening cattle, as profitable to their owners as the mines to theirs; all arable land will be taken up and cultivated, and finally the mountains will be stripped of their timber and will become simply bald and rocky hills. The day when all this shall have taken place is distant no doubt, and will not be seen by the present generation; but it will come.*[39]

IN THE SPRING OF 1880, GRINNELL PASSED HIS DOCTORAL EXAMINATIONS in osteology and vertebrate paleontology. He also completed his thesis on *Geococcyx americanus*—the roadrunner. Somewhere, no doubt, Lucy Audubon was smiling. At the May commencement ceremony, Grinnell's Ph.D. was official. Grinnell, he remembered, had "hoped to work for a long time in the Museum." By the time of his graduation, though, another opportunity already had come his way.

Though *Forest and Stream* was succeeding in its self-declared mission of becoming the preeminent journal for sportsmen of its day, all was not well in the magazine's New York offices. Its high aspirations aside, the journal was not above some good old-fashioned corporate intrigue. According to Grinnell, in the winter of 1879–1880, he was approached by E.R. Wilbur, treasurer of the Forest and Stream Publishing Company. "Mr. Hallock, President of the company, had become more and more eccentric, drinking heavily and neglecting his duties to the paper."[40]

To thicken the plot, Grinnell's father happened to own more than one-third of the stock in the company, and Grinnell himself had "from time to time purchased a few shares." As a practical matter, Grinnell remembered, "we had the control."

At the *Forest and Stream* board meeting of May 1880, Hallock was informed that he would not be reelected. In January 1881, 31-year-old Grinnell assumed both the presidency of the company and the editorial control of the magazine.

A decade of travel to the frontier had imbued in Grinnell deep passion and purpose; now *Forest and Stream* gave him a megaphone on a national stage. The new editor, as he would soon demonstrate, had much to say. For the rest of his life, Grinnell would commit himself to reversing the grim prophecy that he himself had projected for the American West.

CHAPTER NINE

"No Longer a Place for Them"

On this broad continent there is no longer a place for them.

—George Bird Grinnell,
Forest and Stream, *October 18, 1883*

I n 1881 the Northern Pacific Railroad reached the southeastern Montana outpost of Miles City. Like the people of every other town anticipating the arrival of the railroad, the citizenry of Miles City looked forward to a new era of progress and opportunity. "An empire has been brought out of the savage wilderness," said the governor at a ceremony to celebrate the iron horse. Another speaker promised that the railroad would "assuage inequalities of nature and disparities of fortune among our own people, and . . . spread peace, plenty, and prosperity to other nations."[1]

As George Bird Grinnell knew well, the arrival of the railroad was the herald of another certain consequence. In the year before the railroad reached Montana, more than a million buffalo—the northern herd—still walked the continent of North America. Yet the only barrier to their destruction was the remoteness of the territory they occupied. Transportation would surmount that last firebreak, wrote Grinnell, "for as soon as the skins could be brought to a market, the animals that yielded them would be killed."[2]

The railroad pulled full-bore commercial hunting onto the Northern Plains like the caboose on the great train of robber-baron industrialism. In one of his earliest editorials for *Forest and Stream*, Grinnell wrote that "there is nowhere in the world such systematic, business-like and relentless killing as on the buffalo plains."[3]

One of the men who came to Montana was Vic Smith, known as "the champion buffalo hunter." Compared with the men who first hunted buffalo on the Southern Plains, inventing their industry as they went along, the men who came north were skilled craftsmen, artisans of slaughter. They already understood the subtleties of "mesmerizing" a herd through careful selection of shots. And once their quarry was down, northern hunters made sure they profited from what they shot. For ten buffalo killed on the Northern Plains, nine hides now made it to market. Certainly there was incentive for greater care. The high quality of northern hides and the scarcer supply (the southern herd having disappeared) rewarded the hunters' labor with higher prices. Hides now brought as much as $3.50 apiece.[4]

Vic Smith had a leg up on his competition, hunting buffalo in the Yellowstone Valley for three years before the railroad reached Miles City. Smith supplemented his hunting income with a bit of moonlighting. In 1881, he was wanted by the U.S. marshal for smuggling whiskey onto the Sioux reservation. According to Smith, he'd been ratted out by "a mulatto" who was later found dead with a bullet in his brain. Smith casually assumed responsibility for the murder, joking that the cause of the man's death was "lead poisoning." Smith's only punishment, apparently, was the confiscation of his horses.[5]

It didn't keep him from hunting. In the winter of 1881, Smith made his record kill of 107 buffalo in a single stand. By the following year he would face an army of competitors—5,000 buffalo runners swarming every draw, coulee, and butte on the fast-shrinking prairie of Montana, Wyoming, and the Dakotas.

THE EDITORSHIP OF *FOREST AND STREAM* PUT GRINNELL IN A CRITICAL place at a critical time. Defining battles loomed, something Grinnell understood better and earlier than almost anyone. They were battles that would shape the West, and the nation, for all time. At this crucial juncture, *Forest and Stream* provided a weekly connection between Grinnell and a nationwide pool of subscribers, including many of the country's political leaders.[6]

Grinnell fully recognized the potential power of his position, a position in which all of the influences on his early life came to confluence.

In style, Grinnell carried forward the indignant, crusading tone of his predecessor, Charles Hallock. A hard slap of public humiliation continued to be the order of the day for such targets as "trout hogs," "Maine moose murderers," and "law-defying marketmen." Nor did Grinnell hesitate to challenge prominent sporting organizations that he found lacking. Believing that the New York State Association for the Protection of Game, one of the earliest conservation groups, was failing to live up to its legacy, Grinnell challenged it to be more active. "The work we have outlined is no boys' play. . . . Can the Association afford to shirk it?"[7]

Like Hallock, Grinnell applied the velvet glove as often as the iron fist. *Forest and Stream* continued to serve as a clearinghouse for defining good sportsmanship, publicizing recent developments in state and local game protection laws, and educating on natural history (Grinnell maintained his authorship of the natural history page). *Forest and Stream* even attempted to steer its impressionable younger readers away from careers in commercial hunting with articles such as "Our Candid Advice."[8]

While *Forest and Stream* had attempted to influence policy formation from its earliest days—particularly at the state and local level—one of Grinnell's trademarks would be a focus on national politics. In one of his first editorials, "Sportsmen in Congress," Grinnell took note of the fact that the large number of hunters and anglers in Congress provided "another evidence of the constantly increasing popularity of field sports."[9] It was a group on whom Grinnell would soon call.

For political activism at the national level, Grinnell had a mentor in the person of Professor Othniel Marsh. In 1874, while attempting to search for dinosaur bones in the Black Hills, Marsh confronted a furious Chief Red Cloud. Red Cloud, not surprisingly, thought Marsh was after gold. In addition to anger over the invasion of gold-seekers, Red Cloud was also locked in a battle with the Bureau of Indian Affairs over failure to deliver promised annuities to the reservation Sioux. Red Cloud ultimately granted Marsh admittance to the Black Hills, but only after the professor promised to carry Red Cloud's complaints (including samples of spoiled rations) back to Washington.

The senior official responsible for Indian affairs was the Secretary of the Interior, Columbus Delano. Delano's Department of the Interior was notorious for corruption, even by the low standards of the Grant administration. When Secretary Delano brushed back Marsh's inquires contemptuously, the professor began his own investigation. A few years later, Marsh's student George Bird Grinnell would write a glowing profile of Professor Marsh for *Popular Science Monthly*. As part of the article, Grinnell described how Marsh succeeded in exposing the "frauds which he had seen practiced upon the Indians." The Indian Affairs scandal ultimately forced Secretary Delano's resignation. Red Cloud and Marsh became friends, with the Sioux chief even traveling to New Haven and staying at the professor's home. Red Cloud called Marsh the only white man he had ever known who "kept his promises."

The impression of the Indian Affairs incident on Grinnell was profound. "This is perhaps the only instance," he wrote admiringly, "in which a private citizen has successfully fought a department of the Government in his efforts to expose wrong-doing."[10]

Grinnell, like his hero Marsh, was aiming high. After only a month at the helm of *Forest and Stream*, Grinnell wrote an editorial that indicted the very foundation of the Gilded Age. America, he wrote, was "in her money-making stage. . . . The mighty dollar is the controlling agency in every branch of social and public life."[11]

In another editorial, titled "The Very Latest," Grinnell began by noting the great technical advances of the day: "This is an age of progress. . . . We cross the ocean in seven days, the continent in the same space; we talk in our ordinary tones to people miles away, or listen to words that have been stored up in a little box for a week or a year." Grinnell was no Luddite, but he did acknowledge the cost of civilization's advance. "Up to within a few years the valley of the Yellowstone River has been a magnificent hunting ground. . . . The progress of the Northern Pacific Railroad, however, has changed all this. The Indians have been run out and the white men have had a chance to do what they could toward killing off the game."[12]

In the face of an age that celebrated conspicuous consumption, the foundation of Grinnell's emerging philosophy was *self-denial*—the singular lesson he had learned from Lucy Audubon as a boy. Grinnell first applied the principle of self-restraint in the context of his hunting in the

West. The wholesale slaughter of the day contrasts starkly with an incident Grinnell witnessed as he returned from the Yellowstone expedition of 1875:

> *I shall never forget a scene witnessed many years ago, long before railroads penetrated the Northwest. I was floating down the Missouri River in a mackinaw boat, the sun just topping the high bad land bluffs to the east, when a splendid ram stepped out, upon a point far above the water, and stood there outlined against the sky. Motionless, with head thrown back, and in an attitude of attention, he calmly inspected the vessel floating along below him; so beautiful an object amid his wild surroundings, and with his background of brilliant sky, that no hand was stretched out for the rifle, but the boat floated quietly on past him, and out of sight.[13]*

In editorial after editorial, Grinnell called for other hunters to practice the self-restraint that to him was the essence of sportsmanship. Grinnell argued for the outright ban of commercial hunting, as well as tight regulation of sports hunting through the establishment of seasons, bag limits, and licensing fees to generate funds for habitat protection. These are accepted practices today but ran directly counter to the ethics of robber-baron America.[14]

Like Lucy Audubon, Grinnell viewed the primary purpose of self-denial as the protection of future generations. Lucy's focus was intimate—her own family. Grinnell, who married late in life and would have no children of his own, applied the principle more globally—the conservation of natural resources for all generations yet to come. Grinnell invoked the motto of the New York Association for the Protection of Game: *Non nobis solum*. It means "Not for ourselves alone." Hunters, Grinnell argued, were "trustees of the game, to hold and use it for their time and to hand it down to those who are to follow."[15]

For Grinnell, nothing illustrated the self-indulgence he despised more painfully than the extermination of large mammals, the buffalo in particular, in the West. Grinnell used the front page of *Forest and Stream* to berate the president of a group called the Yellowstone Valley Hunting a difference through the establishment of closed seasons. Such laws, how-ever, could work only if enforced, and in this regard, too little—nothing—was done. Patrolling the vast prairie was impractical, but another, quite simple enforcement solution was readily available. Hides could be trans-ported to market only through the chokepoint of railroads. A few marshals in the handful of railheads could easily have halted the trade completely or, at a minimum, enforced closed seasons (winter and summer hides were eas-ily distinguishable). It was not to be.[19]

There occurred one particularly noteworthy attempt to save the buffalo at the federal level. In 1874, Congressman Greenbury Fort of Illinois brought a bill to the floor of the U.S. House of Representatives that would have prohibited the killing of female buffalo in any season and prohibited the killing of male buffalo except for food. Indians were excepted from the law. Penalties for violation included a fine on the first offense ($100 per buffalo killed), and on a second offense, imprisonment.[20]

There followed a lengthy debate, both stunning and telling. Repre-sentative Samuel L. Cox of New York objected strenuously to the bill's exception for Indians, citing the fact that "The Secretary of the Interior [Columbus Delano] has already said to this House that the civilization of the Indian is impossible while the buffalo remains on the plains."[21]

Congressman Fort responded sharply. "I cannot understand why the Secretary of the Interior should have used this language to the gentle-man or to his committee, but certainly as an individual I am not in favor of civilizing the Indian by starving him to death, by destroying the means which God has given him for his support."[22]

The debate continued, with Congressman Omar Conger of Michi-gan complaining that the buffalo "eat the grass. They trample upon the plains upon which our settlers desire to herd their cattle and their sheep. . . . They are as uncivilized as the Indian."[23]

Representative Isaac C. Parker of Missouri attempted to offer a more nuanced interpretation of Secretary Delano's view on buffalo–Indian policy:

> *In my judgment, the great key to the solution of this Indian problem is to confine these Indians upon as small a tract of land as possible, and if possible to make it a necessity for them to*

Club. Grinnell took no issue with the idea of hunting itself, a pasti
he held nearly sacred. But he strongly protested the president's perpe
ation of the myth of inexhaustibility, or as Grinnell described it,
claim that "the game is so abundant that it cannot be exterminated."

"[L]et us tell the president that he is wrong. . . . It is but about e
years since we could see along the Platte River and thence south to
Republican, and beyond through Kansas, the Nation and Texas, buffal
thousands and hundreds of thousands. . . . Where are they now?" Grir
warned explicitly of what was to come. The buffalo "will disappear f
the Valley of the Yellowstone, unless steps are taken to protect it there

As editor, Grinnell also publicized the writings of others who warn
the buffalo's demise. In 1881, Grinnell ran a long letter by a reader iden
only as "C.B." Under the headline "Buffalo Extermination," C.B. ar
that imminent demise of the buffalo was "abundantly evident . . . the
consequence of the merciless war which has been waged against t
C.B. also directly addressed an issue that stood at the heart of the matt
conflict between preserving the buffalo and preserving the jobs of hu
"Even if legislation should compel some men to seek another mc
making a living, it would be better than to exterminate the buffalo.
called on the federal government to take action. "Cannot somethi
done to put a stop to this wanton destruction? Is there not some
which could be taken on the part of the General Government that
do something toward the preservation of the bison?"[17]

Other *Forest and Stream* readers also called for legislative action, i
ing one from Helena, Montana. "I do hope Congress will pass a
protect our buffalo and other large game, now being slaughte
wholesale. It is reported that one man on our border killed 2,000
this winter for their hides only. At that rate how long can it last?"

LEGISLATIVE EFFORTS TO PROTECT THE BUFFALO, CIRCA 188
into two categories: too late and too little. In the "too late" c
came state and territorial actions by Colorado, Kansas, Nebras
the Dakotas. Each of these enacted tough-sounding laws to pro
buffalo only after the animal had been exterminated within its l

In the "too little" category came actions by Idaho, Wyoming, an
tana. These territories passed laws early enough that they might ha

learn to labor and to get a sustenance from the soil as the white man does, and not depend upon the rivers and the plains to furnish them their fish and their game. That is the reason why the Secretary of the Interior entertains this opinion.[24]

Parker's refinement failed to impress Congressman Fort, and it was he who would have the final, sarcastic word: "Shoot the buffalo, starve the Indian to death, and thereby civilize him! I would suggest that a shorter and more humane way would be to go out and shoot the Indians themselves—put an end to their existence at once, instead of starving them to death in this manner."[25]

Both the House and the Senate would vote to protect the buffalo, sending a bill to President Grant for signature into law. Sadly for the buffalo, the Congress adjourned immediately after passage of the protection legislation. Adjournment opened the door for killing the bill with no fingerprints—by pocket veto. The bill sat on Grant's desk for the requisite ten days with no signature, and after that it was dead, never to be revived.[26]

It is not clear whether Secretary Delano intervened directly to urge Grant against signing the bill, but Delano's support for the deliberate extermination of the buffalo had been a matter of public record even before Congress took up the 1874 bill. In his annual report of 1873, Secretary Delano wrote that "I would not seriously regret the total disappearance of the buffalo from our western prairies, in its effect upon the Indians."[27]

In the back rooms of Washington, D.C., it was Delano's doctrine that had prevailed. And in the space of less than ten years, it would prevail on the plains as well.

THE CANADIAN PORTION OF THE NORTHERN HERD WAS THE FIRST TO succumb. Commercial hunting of the buffalo, in fact, could trace its earliest roots to Canada. The first record of large-scale buffalo hunting took place in Manitoba in 1820, when a caravan of 540 carts rolled out to gather up the buffalo harvest on the Canadian plains. Nevertheless, for many decades, physical isolation afforded the animals a significant measure of protection. In the United States, the express policy of manifest destiny encouraged westward migration. In western Canada, though, the omnipotent Hudson's Bay Company successfully discouraged settlement, and even the development of transportation lines, to protect its monopoly on fur resources. Still, by the 1870s, Canadian hunters were using roads leading south to transport hides into the American market.[28]

The coup de grâce for the Canadian herd came in the aftermath of the 1876 Battle of the Little Bighorn, when Sitting Bull and a large band of his followers sought refuge in the land of the "Great Mother," Queen Victoria. In an 1878 effort to starve the Sioux into surrender—and settlement on U.S. reservations—the U.S. Army began concerted efforts to block the northward migration of buffalo into Canada. In addition to burning great stretches of borderland grasslands, the United States also allegedly deployed a "cordon of half-breeds, Indians, and American soldiers" to haze the buffalo away from Canadian territory. American activity on the border would prompt official protest by the Canadian government. The United States denied a policy of "defoliation."[29]

Whether U.S. actions were undertaken as a matter of formal policy or merely local practice, they ultimately would succeed. Combined with commercial hunting and subsistence hunting by Canadian Indians and American Indian refugees—the blockage of north–south migration routes spelled the end for the Canadian herd. By 1880, it was gone.[30] By 1881, Sitting Bull would surrender his starving band in the Dakota Territory, asking history to remember that he was "the last man of my tribe to surrender my rifle."[31] The last free Indians of America's Northern Plains had capitulated to reservation life.

For the Montana herd, the beginning of the end was Miles City, temporary terminus of the Northern Pacific Railway. Montana buffalo runners estimated that half a million buffalo were cut down within a 100-mile radius of Miles City, called the Dodge City of the northern hunt.[32] Vic Smith made his record kill near the railhead, and in 1881 he claimed a total harvest of about 4,500.

Montana's harsh winters were an ally to commercial hunters. Not only did the frigid temperatures improve the quality of hides but they also caused the buffalo to bunch together in the few valleys where pasturage could be found. According to a piece in the *Sioux City Journal,* "There was no sport about it, simply shooting down the famine-tamed animals as cattle might be shot down in a barn-yard."[33]

A skinner strips hides from the fast-dwindling members of the northern herd, circa 1882.
Courtesy of Archives and Special Collections, University of Montana-Missoula.

Prior to the arrival of the railroad, winter had been the only season in which the northern herd was pursued by commercial hunters. The Missouri River trade had been in robes—treated hides. Robes could be harvested only in winter, when frigid weather made the pelage velvety plush. The focus on robes afforded the buffalo many months of reprieve, including the critical spring–summer calving season. With the arrival of the railroad, however, hunters now also pursued the buffalo for hides—the untanned product that was processed into leather. As on the Southern Plains of the 1870s, northern buffalo runners could now hunt year-round.[34]

By 1882, the center of the great hunt had shifted to a vast expanse of plains at the heart of the Montana Territory. Hunting camps hemmed the banks of three rivers—the Missouri, the Yellowstone, and the Musselshell—that formed a loose triangle around the last great killing fields. The camps, according to an army officer in Montana named Lieutenant Partello, rendered it "impossible for scarcely a single bison to escape through."[35]

None did. The most notable event in the winter of 1882–1883 was the panicked attempt of a herd estimated at 75,000 to push north across the Yellowstone River. In Lieutenant Partello's description, the buffalo were set on by a swarming mass of "Indians, pot-hunters, and white butchers," shot down as they emerged from the river, then hounded as they fled north.[36]

A similar incident finished off the buffalo in the Dakotas. In the fall of 1882, rumors began to circulate the Dakota plains that a herd had been sighted on the expanse of prairie between the Black Hills and Bismarck. No meaningful numbers of buffalo had been seen in this territory for years, and the news seemed almost miraculous to the Sioux Indians of the Standing Rock Reservation, the new home of Sitting Bull.

The agent at Standing Rock gave Sitting Bull and 1,000 warriors permission to leave the reservation for a hunt. His rationale was pragmatic, not sentimental. Food rations at Standing Rock were short, and

THE LAST BUFFALO.

"Don't shoot, my good fellow! Here, take my 'robe,' save your ammunition, and let me go in peace."

A period cartoon reflects the dawning political concern over the destruction of the buffalo.

Courtesy of the National Park Service, Yellowstone National Park.

without the supplement of buffalo, the agent worried about his ability to carry the tribe through the long winter to come.[37]

White hunters also caught scent of the herd, including Vic Smith and his self-described "relentless pursuit of the last of the noble bison." The hunt took on a circuslike air, complete with rumors that Buffalo Bill himself might show up. So it was that a mixed group of whites and Indians fell upon the last herd of Dakota buffalo. "[W]hen we got through the hunt," said Vic Smith, "there was not a hoof left." Even then they called it "the last Indian buffalo chase." In the Sioux camp after the hunt, Sitting Bull reportedly spent much of his time sitting apart from the others, alone to contemplate the stunning events of his life.[38]

In the fall of 1883, thousands of Montana buffalo hunters geared up for the winter hunt. They purchased ammunition, provisions, and mules, then rolled out across the "great triangle" to stake their turf. But the northern herd was gone. Not gone to Canada, as many surmised. Gone forever as wild animals on the plains.

The numbers paint the stark picture at the end. In 1882, the Northern Pacific Railroad alone shipped 200,000 hides to eastern processing facilities, an amount that filled an estimated 700 boxcars. In 1883, the railroad shipped 40,000 hides. In 1884, the total harvest fit in a single boxcar, and according to a Northern Pacific official, "it was the last shipment ever made."[39]

IN 1885, WARNER, BEERS & CO. PUBLISHED A 1,300-PAGE *HISTORY OF Montana* (which seems all the more weighty given that Montana was still four years away from statehood). In a lengthy section on the territory's natural history, the editors described "wild animals in large numbers." Despite the "continuous warfare waged against them by both Red and White hunters," wrote the editors, "representatives of the different classes are still to be found in the forests and mountain fastnesses of the territory." As for the buffalo, though, *History of Montana* called it "almost an animal of the past."[40]

Certainly commercial hide hunting, by 1885, was an industry of the past. With the end of a steady supply, leather manufacturers turned back to cattle. Ranching enterprises flooded the void prairie once filled by Indians and buffalo.

Cowboys still stumbled across scattered buffalo, and even a few larger groups, for a number of years to follow. When they did, their instinctive action was usually a hand "stretched out for the rifle." Ed Carlson, a Montana cowboy, encountered a small group of seven buffalo while riding the range in 1885. He blazed away until his gun was empty and succeeded in killing one. While Carlson struggled to reload, a massive old bull spun around to confront him at a distance of "not over twenty-five yards." The bull stood "perfectly still," remembered Carlson, "seeming to say in his mute way: 'I am the last of my race; shoot me down.'" Then it scampered away. "That was the last wild buffalo I ever saw on the plains."[41]

Rich easterners came east to hunt the buffalo before they were gone, including a young New Yorker named Theodore Roosevelt. The primary motivation of the 24-year-old Roosevelt's first voyage to the Dakotas—a trip that changed the course of his life—was the desire to kill a buffalo. For his guide, Roosevelt selected an experienced companion—Vic Smith. A painting from the 1890s, *Theodore Roosevelt and His Hunting Guide, Vic G. Smith*, shows the two men on galloping horses, Roosevelt closing for the kill.[42]

All over the West, hunters vied for the distinction of "killing the last wild buffalo." One man who watched the "alarming progress" of the bison's extermination was William Temple Hornaday, chief taxidermist at the Smithsonian Institute. By 1886, according to Hornaday, the killing of a buffalo had become so rare an event as to be "immediately chronicled by the Associated Press and telegraphed all over the country." Against this backdrop, Hornaday fretted that "the Museum was actually without presentable specimens of this most important and interesting animal." So Hornaday set out to kill some. In fact, he hoped to kill twenty.[43]

That Hornaday did not reflect a bit more on his actions seems startling today, but it paints an accurate picture of the degree to which conventional wisdom assumed that the complete extinction of the buffalo was both imminent and unavoidable. Hornaday, to his everlasting credit, would later emerge as one of the (living) buffalo's most important defenders.

In 1886, though, Hornaday's focus was on collecting specimens for the museum. In September he arrived in Miles City at the head of a well-financed expedition. They hunted for three months on Big Dry Creek, a remote tributary of the Missouri whose extreme terrain gave cover to a few scattered survivors. In a breathless, twelve-page description of the hunt, Hornaday shows few hints of his future as a conservationist. When it became clear that the expedition would likely reach its goal of killing twenty buffalo, Hornaday wondered about "the possibilities of getting thirty." They would ultimately killed twenty-five, only six of which would be mounted for display.[44]

Nor did Hornaday demonstrate particular insight to the plight or culture of the Indian. At one point during the three-month hunt, members of Hornaday's party killed a large bull that they field-dressed, then left on the plains to be collected in the morning. Upon returning to the site, they discovered that "a gang of Indians had robbed us of our hard-earned spoil." The Indians, complained Hornaday, had "stolen the skin and all the eatable meat, broken up the leg-bones to get at the marrow, and even cut out the tongue." Hauntingly, the Indians took one other action with the dead bull. As Hornaday described it, "[T]o injury the skulking thieves had added insult. Through laziness they had left the head unskinned, but on one side of it they had smeared the hair with red war-paint, the other side they had daubed with yellow, and around the base of one horn they had tied a strip of red flannel as a signal of defiance."[45] What Hornaday interpreted as personal slight aimed at the white hunters seems clearly to have been a ritual of respect toward the buffalo.

Hornaday himself would shoot the last buffalo of the Smithsonian hunt—a massive 1,800-pound bull that would stand as the centerpiece of the National Museum of Natural History diorama. When Hornaday skinned his bull, he made a discovery. "Nearly every adult bull we took carried old bullets in his body, and from this one we took four of various sizes that had been fired into him on various occasions. One was sticking fast in one of the lumbar vertebrae."[46]

Hornaday spent a year mounting the six buffalo that came to be known as the Hornaday Bison Group. His technique was revolutionary for its time, presenting the animals in natural poses and even replicating the look of their prairie setting. The display was a popular sensation, an icon of the Smithsonian until its removal in 1957. Through his diorama, Hornaday believed he had given the buffalo something they could not otherwise achieve—immortality. "Perhaps you think a wild animal has no soul," he wrote, "but let me tell you it has. Its skin is its soul, and when mounted by skillful hands, it becomes comparatively immortal."[47]

The Hornaday Bison Group: Once a centerpiece of the Smithsonian Institution in Washington, D.C., Hornaday's buffalo are now on display at the Museum of the Northern Great Plains in Fort Benton, Montana. *Courtesy of the Overholser Historical Research Center, Fort Benton.*

GEORGE BIRD GRINNELL SHARED IN PART OF THE CONVENTIONAL wisdom of the day concerning the future of the buffalo. The press of civilization, he believed, was inexorable, and "while it may be deplored, it cannot be avoided." But unlike most of his contemporaries, Grinnell held out one hope for survival of the buffalo as something other than a display in museums.

As early as 1877, even before the final destruction of the southern herd, Grinnell had advocated the establishment of what he called "a buffalo reservation." Having recently explored the upper Yellowstone, Grinnell saw in the nation's first national park the potential "for saving the few remaining herds from utter annihilation." In Yellowstone, he wrote, "we have the necessary territory, and it is already stocked." But as Grinnell had witnessed for himself, saving the buffalo would require defeat of a determined foe: "[T]he skin-hunter, that ruthless destroyer of game, must be kept at a distance, if we would hope to save this species."[48]

For most of the Montana skin-hunters who geared up in the fall of 1883, dawning awareness of the complete destruction of the northern herd marked the end of their hunting careers. They sold the tools of their trade at salvage rates and got out. Even the revered Sharps buffalo rifle saw its market value plummet—too much gun for other game. Most of the hunters moved into ranching, Montana's newest industry.[49] Others drifted to Butte, where a copper boom (the nation had recently gone electric) created constant demand for miners. For a few men, though, hunting ran deep in their blood.

The supply of buffalo would no longer support the leather industry, but the law of supply and demand remained in full force. A new demand arose from the very scarcity of the buffalo—a wistful nostalgia for a Wild West that by the mid-1880s already had disappeared. Suddenly every saloon in America (and many in Europe) wanted a mounted buffalo head to hang on the wall—one last frontier trophy. And if the number of living buffalo was now only a few thousand, that just meant that the price went higher. In Livingston, Montana, near the northern entrance of Yellowstone National Park, taxidermists paid poachers up to $500 per head for buffalo that only a few years before had blackened the plains.[50]

Vic Smith, for one, smelled new opportunity. Leaving the Dakotas behind, he packed up his rifles and headed for Yellowstone.

As Grinnell realized, the survival of the buffalo as a species was now inextricably linked with the outcome of a haphazard, neglected experiment: successful protection of Yellowstone, the nation's first national park.

CHAPTER TEN

"Blundering, Plundering"

. . . noisy, ever changing management, or mismanagement,
of blundering, plundering, money-making vote-sellers who
receive their places from boss politicians as purchased goods . . .

—JOHN MUIR,
summarizing early civilian administration
of Yellowstone National Park[1]

A mid-1880s scandal paints a telling portrait of the early administration of America's first national park. In 1882, through adoption of a treaty, Congress extinguished the title of the Crow Indians to a parcel of land on the northeastern border of Yellowstone. Prospectors had invaded the territory even prior to the new treaty, but with title settled definitively, the region drew the interest of East Coast investors. One of them was none other than Jay Cooke, the railroad financier whose bankruptcy had plunged the nation into the Panic of 1873.

Cooke toured a remote camp dubbed Shoo-fly near the northeastern border of Yellowstone Park. He liked what he found, and promised the camp's 150 resident miners that he would invest in their town. Most significantly, he promised to bring them the railroad. The delighted miners rechristened their camp Cooke City.[2]

As an engineering feat, connecting Cooke City to the Northern Pacific Railroad was a small matter. Less than sixty miles of broad valley separated the mining town from the nearest connecting point to the railroad. There was, however, a *legal* impediment: the valley, vital range for large herds of elk and buffalo, ran through Yellowstone National Park. Laying rails, therefore, would require an act of Congress. Jay Cooke, the Northern Pacific Railroad, and an army of lobbyists set to work.

In the fall of 1884, President Chester A. Arthur named a new superintendent to Yellowstone Park. His pick was Robert E. Carpenter, a political appointee from Ohio. One of Carpenter's promising initial acts as chief executive was to clear out illegal squatters from the northeastern corner of the park—an action that won guarded praise from George Bird Grinnell on the editorial pages of *Forest and Stream*. More generally, though, Grinnell withheld judgment: "The new Superintendent of the Park will have an opportunity during the season that is coming to show what stuff he is made of. He may be sure that his actions will be scrutinized closely. If he does his duty he will be applauded, but if he fails it will soon be known."[3]

In the winter of 1884–1885, Superintendent Carpenter was not acting like a man under scrutiny. Leaving the park behind, the superintendent traveled to Washington and took up residence with Carroll T. Hobart, a man in the midst of his own scandal over illegal use of park resources. Together with a coterie of railroad lobbyists, Hobart and Carpenter spent the winter lobbying Congress in support of a policy called segregation. In the context of the 1880s, segregation meant carving off a portion of Yellowstone National Park and returning it to private hands—with the primary goal of building Jay Cooke's railroad connection to the mines.

Superintendent Carpenter knew that segregation, if it became law, would result in a land rush to claim the formerly protected properties. Though he had been in Yellowstone for only a few months, the superintendent had taken notice of several prime locations. At the same time he was clearing out squatters—who would hold competing claims if the lands became available—Carpenter was scouting choice sites, including an area on Mount Evarts thought to contain large deposits of coal.

By February 20, 1885, passage of the segregation bill seemed certain, and Superintendent Carpenter proceeded with a plan that Grinnell would later call "either the height of audacity or a most dense ignorance." Carpenter first sent a secret telegraph to a coconspirator, his son, in Livingston, Montana, a railroad town fifty-five miles from the park.

Carpenter's son then placed a telephone call to another cohort, awaiting word at a semipublic phone in a general store at Gardiner—on the park's northern border. To avoid tipping off the locals who congregated in the store, the plotters had prearranged a code. On hearing a coded phrase, the men in Gardiner were to rush out and stake claims on the choice ground.[4]

Unfortunately for Carpenter's ham-handed plot, telephone service between Livingston and Gardiner was less than crystal clear. The first coded phrase, "Secure the horse at once," could not be understood by the man in the Gardiner general store, leading him to shout loudly into the phone for clarification. At the Livingston end, the caller switched to a backup code: "No wind in Livingston." By the time this phrase had been clearly conveyed, it had been broadcast to everyone in the store—about twenty people—leading to much suspicion that something was afoot. The notorious Livingston winds blow twenty-four hours a day, and in the words of a local newspaper editor, the phrase "no wind in Livingston" was "so manifestly mysterious and doubtful in its probability that it aroused suspicion outside the circle for whose benefit it was particularly intended."[5]

At dawn the next morning, as Carpenter's gang prepared to ride out of Gardiner and stake claims, they found themselves in the company of "a stampede party of forty or fifty persons" now more or less privy to the plot. Carpenter's men still managed to reach their preselected sites, driving location stakes in prime property that included 1,400 acres with the name "R.E. Carpenter" attached.[6]

There was only one problem: To Superintendent Carpenter's great shock, the segregation bill failed to pass. For the time being, anyway, his would-be coal mine would remain part of Yellowstone National Park. In Cooke City, meanwhile, the newspaper got wind of the fact that Carpenter—who had so recently been clearing out squatters—then staked his own claims. A local outcry erupted, soon picked up by the watchful editor of *Forest and Stream*.

Under the headline "Remove the Superintendent" on April 9, 1885, Grinnell charged that Carpenter was "pecuniarily interested with the Hobarts in various projects which depend for their success on the reduction of the size of the reservation, and that he has already laid claim to a portion of the public domain within the Park." Two weeks later, Grinnell accused the superintendent of having "enrolled himself among the number of the land-grabbers who have endeavored to wrest from the people a portion of their pleasure ground."[7]

Carpenter sputtered out denials for a few months, but by June he was fired. Grinnell had won a battle, but the war over Yellowstone and its resources was a decade away from its dramatic finale.

CONCERN ABOUT COMMERCIAL EXPLOITATION OF YELLOWSTONE's natural resources had been the primary driving force behind the 1872 establishment of the region as a national park, at least in theory. Despite the supposed centrality of this purpose, however, Congress's general indifference to Yellowstone's management opened the floodgates to all manner of moneymaking schemes, including poaching.

Commercial opportunists, even beyond the trappers and miners, were an early and then ever-present fixture in the upper Yellowstone. When the 1871 Hayden expedition tramped up the Gardner River to Mammoth Hot Springs, they were shocked to find an "impromptu village," a place dubbed Chestnutville in honor of its proprietor, one Colonel Chestnut. The colonel was operating a crude spa for customers who swore by Mammoth's steaming waters as a curative for rheumatism. Reputedly, the hot mineral springs also provided relief for another common frontier ailment—syphilis. (Several early plans for Yellowstone sketched grand visions for the Park's waters, including an inane 1874 proposal to create a "national swimming school" and a rowing club.)[8]

The act of Congress creating Yellowstone gave "exclusive control" of the national park to the Secretary of the Interior, who was charged with making rules and regulations "for the preservation, from injury or spoliation, of all timber, mineral deposits, natural curiosities, or wonders within said park, and their retention in their natural condition." The act specifically addressed wildlife, directing the secretary to "provide against the wanton destruction of the fish and game found within said park, and against their capture or destruction for the purposes of merchandise or profit."[9]

The power of this law, seemingly unambiguous on paper, quickly degenerated from the vantage of the park itself. Yellowstone was a legal void—federal property but also part of the territories of Montana and Wyoming (which would not become states until 1889 and 1890, respectively). The law creating Yellowstone established no mechanism for government or law enforcement in the park. With no precedent, there was no clear idea of what it meant *to be* a national park. As a result, administration of Yellowstone in its first decade and a half of existence (with a few notable exceptions) ranged between passive neglect and active exploitation.

To serve as Yellowstone's first superintendent, the Secretary of the Interior picked Nathaniel Langford, the man whose 1870 expedition helped to launch the effort to create the park. Though a bit of self-promoter, Langford's devotion to the park ideal was sterling. Langford learned of his appointment while in Cheyenne, Wyoming. In accepting the new position, he sent a telegraph to the secretary and asked to be advised of "what power my appointment gives me."[10] Whether Langford ever received a reply is unclear, but he would soon discover for himself the limits of both his authority and Congress's interest in the new national park.

For starters, Congress had declined to appropriate any money for the park, so Langford learned that he would serve without salary. Indeed Yellowstone was an early example of what today might be called an unfunded mandate. In creating the park, Congress had stated lofty "objects and purposes." As to how these aspirations should be realized, Congress had little interest.

Langford drafted the park's first regulations, including the following: "All hunting, fishing, or trapping within the limits of the Park, *except for purposes of recreation, or to supply food for visitors or actual residents*, is strictly prohibited, and no sales of fish or game taken within the park shall be made outside of its boundaries."[11]

In his official report of 1872, Superintendent Langford recommended the establishment of "severe penalties" for breaking park rules. Washington ignored the suggestion. Langford even went so far as to suggest attaching the park to Montana so that that territory's laws and officials could be enlisted in enforcement efforts.[12] Wyoming, whose borders contained 98 percent of the park, took a dim view of this proposal, and the two territories succeeded in canceling each other out.

The park's second administrator was a tough-minded man of action named Philetus W. Norris. Though a resident of Norris, Michigan (a town named in his honor, now a part of Detroit), Norris had spent con-

Philetus W. Norris: Norris's garb was appropriate for his man-of-action approach to administering Yellowstone, but congressional indifference limited his effectiveness.
Courtesy of the National Park Service, Yellowstone National Park.

siderable time in the West. In fact, Norris was cut from the same cloth as his good friend—Charlie Reynolds. On Norris's voyage to assume his duties in the park, he made a pilgrimage to the Little Bighorn battle-field, collecting Reynolds's body for reinterment.[13]

As Norris traveled toward his posting in Yellowstone, he carried with him 500 copies of "spirited cautions against fire and depredations in the park." The notices were printed on cloth and nailed to trees in and around the park, warning visitors against activities including the "wanton slaughter of rare and valuable animals." The new superintendent invoked the "good fame of the people of Montana, Utah, and Wyoming." Norris, who patrolled Yellowstone in fringe-trimmed buckskin, hoped his handbills would have a particular impact on his "old mountain comrades and friends." But these "mountaineers," as Norris called them, bluntly told the new superintendent how things worked: So long as the government did nothing to protect the game, they would continue to hunt it.[14]

Most of the park, Norris believed, was by 1877 already "too much decimated to justify extra efforts for its protection." In some places, though, he still held out hope. The superintendent gave as a specific example a herd of 300 or 400 buffalo in the northeastern portion of the Park near Cooke City. As for how to protect these few remaining vestiges, he called for "manly treatment and proper rules uniformly enforced."[15]

In 1878, six years after the establishment of Yellowstone National Park, Norris succeeded in obtaining its first appropriation from Congress—the decidedly modest sum of $10,000. Norris became the first Yellowstone superintendent to receive a salary and also was able to hire a few assistants to aid in patrolling the park's 2.2 million acres. Not all of them worked out to his satisfaction: He later had to reprimand one assistant, J.C. McCartney, who, while still in his official capacity, established a residence in the park for the purpose of selling spirits.[16]

Another Norris hire was more faithful to the spirit of his duty. Norris hired the park's first gamekeeper, a crotchety old dog named Harry Yount. According to Yount's 1880 report, he built a sturdy cabin in the game-rich Soda Butte Valley outside of Cooke City and wintered there "so as to protect the game, especially elk and bison, in their sheltered chosen winter haunts." Yount, though, admitted frankly the limits of his own effectiveness. The job, he wrote, "cannot be done by any one man, and I would respectfully urge for the purpose the appointment of a small, active, reliable police force."[17]

Norris served as superintendent until 1882. The Northern Pacific Railroad reached Livingston the same year, creating, for the first time, the prospect of easy access to Yellowstone. Since the establishment of the park, the government had hoped that leases to hotel vendors would fund the construction of roads. Now, finally, with the prospect of visitors delivered by rail, the moment seemed to have arrived. A New York–

Harry Yount: In 1880, Yellowstone's first game warden reported that his job could "not be done by any one man, and I would respectfully urge for the purpose the appointment of a small, active, reliable police force." *Courtesy of the National Park Service, Yellowstone National Park.*

capitalized entity calling itself the Yellowstone Park Improvement Company promised to throw money at a major project to construct hotels and other visitor comforts. At *Forest and Stream*, Grinnell's fine-tuned senses caught a whiff of the malodorous.

In an editorial on January 4, 1883, titled "The Park Grab," Grinnell told his readers how the improvement company officials "related with tears in their eyes most heartrending stories of the slaughter of game, and told about the destruction of geysers and other natural wonders." For people "ignorant to the subject," Grinnell continued, the officials made it appear "that they were acting almost entirely from philanthropic motives." As the terms of the proposed deal made clear, this was no act of charity. For an annual rent of $2 an acre, the improvement company would receive 640 acres (one square mile) around each of Yellowstone's seven biggest attractions. To this exclusive dominion over major sites would be added a government-granted monopoly for stagecoach transportation, telegram communications, management of stores, guiding services, and even boating on Lake Yellowstone. To build its facilities, moreover, the company would have the right to cut as much park timber as it needed.[18]

A week later, Grinnell informed his readership of another improvement company activity in the park. Despite the fact that no contract had yet been formally agreed on, timber cutting and full-scale construction of a hotel was proceeding at Mammoth. To feed the 100 men building the hotel, the company had contracted with local hunters to supply 20,000 pounds of meat. Rather than expend the funds to import beef, the improvement company would save money by feeding its army of workers with "elk, deer, mountain sheep, and bison, killed in the park."[19]

In fighting against the Yellowstone Park Improvement Company, Grinnell honed an argument that would be central to all of his future

Commercialization of Yellowstone: A *Harper's Weekly* cartoonist lampoons a tourist's arrival at the park in the January 20, 1883 edition.

conservation efforts: Under a headline of "The People's Park," Grinnell called on Congress to protect the "interest of the people . . . from all monopolistic land-grabbing schemers." Yellowstone, as Grinnell explained, belonged to everyone. "Every citizen shares with all the others the ownership in the wonders of our National pleasure ground, and when its natural features are defaced, its forests destroyed, and its game butchered, each one is injured by being robbed of so much that belongs to him." In making this case, week after week, eventually year after year, Grinnell was defining the meaning of such novel concepts as "public land" and "national park."[20]

"It is the duty of this journal," wrote Grinnell, "to do all in its power to protect the interests of the people, to guard against any invasion of their rights, and to sound the note of warning and alarm when these rights are threatened." Grinnell told his readers that they had a personal stake in the outcome of the fight for the park. For most Americans, Yellowstone might be distant and remote—but it was *theirs*. This was no European-style reserve for the noble and rich. "The Park is for rich and poor alike," said one Grinnell editorial, "and every one should have an equal interest in it."[21]

By the end of January, the Department of the Interior was feeling the heat. In a letter to Senator George Vest of Missouri—Grinnell's closest ally on Capitol Hill—Secretary of the Interior Henry Teller put some distance between himself and the Yellowstone Park Improvement Company. The secretary promised that until Vest and the rest of Congress gave more guidance (Vest was a known opponent of the improvement company), he would "take no action in the matter of the lease." In response to concerns about the improvement company's killing of park wildlife, Secretary Teller also issued a new regulation "prohibiting the killing of game within the park." This at last closed the gaping legal loophole that had allowed killing "for purposes of recreation, or to supply food for visitors or actual residents."[22]

"This ends the fight," declared Grinnell in a self-congratulatory editorial titled "The Park Saved." "The grabbers are defeated. The people's rights are to be protected. The Yellowstone Park is not to become a second and greater Niagara."[23] For the normally clear-eyed Grinnell, it was a rare instance of overexuberance.

As Grinnell knew better than anyone, the creation of rules was meaningless without enforcement. Indeed the importance of enforcement had been one of Grinnell's signature issues since assuming the editorship of *Forest and Stream*. "[T]he main trouble has been not so much in the law itself," he argued in one early editorial, "as in its lack of enforcement." As for Yellowstone, Grinnell lambasted the federal government for having "again stultified itself by enacting laws without supplying the means to enforce them." In another piece, he argued that "Rules and regulations are all very well, but there must be some power to enforce the rules and regulations which have been made."[24]

The federal act creating Yellowstone, combined with Department of the Interior regulations, already protected park resources in theory. With the Department of the Interior's closure of the loophole that allowed hunting for recreation and subsistence, there could be no ambiguity about the illegality of hunting within park boundaries. Yet all variety of carnage and plundering continued. According to Hiram Chittenden, who wrote the first comprehensive history of Yellowstone in 1895, the first two decades of the park witnessed "evils of a license which at times was wholly unchecked."[25]

Tourists, as Grinnell observed firsthand in 1875, continued to hack up slabs of geyserite for souvenirs and to carve their names into all available surfaces. Another popular activity, according to one assistant superintendent, was to "test the power of the Geysers by throwing timber in them." Then there was the perennial problem of forest fires sparked by carelessly tended campfires. General Philip Sheridan and his men, touring the park in 1882, accidentally set off a conflagration that scorched 15,000 acres. Yellowstone's forests also fell victim to illegal logging. Individuals from border communities such as Gardiner and Cooke City hauled out firewood by the wagonload for winter fuel.[26]

The destruction of wildlife that Grinnell witnessed in 1875 had sparked a lifetime of advocacy in support of conservation. A decade later, even after new regulations prohibiting all hunting in the park, the slaughter of game continued unabated. Many tourists, in the great "plinking" tradition, treated the park like a giant shooting gallery. A

"Hunting Buffalo, Yellowstone Park": A nineteenth-century stereoscope slide gave viewers a three-dimensional image of one of Yellowstone's early tourist attractions.
Courtesy of the National Park Service, Yellowstone National Park.

stereoscope card from the Cosmopolitan Series showed a hunter beside a dead buffalo with the caption "Buffalo Hunting, Yellowstone Park." Tourists shot whatever moved: deer and elk, badgers and prairie dogs, geese and swans. One man boasted of coming on six bears and killing five of them. With the frequency of such salvos, it is hardly surprising that another early visitor wrote of seeing elk carcasses on "every hillside and in every valley." As for live game, he wrote that "it is a rare thing now to see an elk, deer, or mountain sheep along the regular trail from Ellis to the Yellowstone Lake."[27]

Hunting by tourists caused a broad retreat of animals from the park's main travel corridors, but it was commercial hunters—and commercial fishers—who posed the greatest threat to Yellowstone wildlife. One common source of food for Cooke City's hungry miners was Yellowstone's Fish Lake (now Trout Lake). Fishermen dredged large quantities of trout from the lake through the use of spears, nets, and dynamite— tossing the explosive into the water, then scooping up the stunned fish when they floated up to the surface.[28]

Commercial hunters also employed tactics of systematic extermination, scouring the park's most distant corners. Elk, a common target, brought $6 a hide. But it was the buffalo that represented the biggest prize. With buffalo selling at hundreds of dollars for a single head, one lucky hunt could provide a poacher with income equal (for comparison) to a year's wages in Butte's dangerous underground mines.

To such bounteous financial incentive could be added the miniscule threat of detection and capture. In the first decade of the park's history, there was no organized police force to patrol Yellowstone's 2.2 million acres. In the earliest years, there were extended periods when no federal official whatsoever was even present in the park. Even after that first apportionment of $10,000, one or two men shared responsibility for patrolling an area 50 percent larger than the state of Delaware.

In 1883, Congress finally established a modest appropriation for the first semblance of a police force. The Department of the Interior was given funds to hire ten "assistant superintendents" at a salary of $900 per year. Grinnell used *Forest and Stream* to warn the Yellowstone superintendent that "if he appoints as his assistants a lot of Eastern men who know nothing of the mountains and the habits of game, the old style slaughter will be kept up." Instead, the superintendent should "make his selection of these officers from among the mountain men of Montana," who would be "faithful and fearless in the performance of their duty." The Department of the Interior ignored Grinnell's advice. Instead of selecting well-qualified frontiersmen, experienced and equipped to patrol Yellowstone's harsh backcountry, the appointments were doled out as political patronage.[29]

The result was a group of East Coast tenderfeet whom the local Montanans derided as "rabbit catchers." One unimpressed observer of the new crew thought that "a prairie wolf would frighten them out of their pants." Another believed that a "couple of cowboys could put the whole brigade to flight with blank cartridges."[30]

The poachers that the superintendents faced, meanwhile, were the flinty, antiauthoritarian products of buffalo hunting, mining camps, and railroad crews. And they knew the score. When an assistant superintendent named Jason Dean attempted to enforce park rules, he was "laughed at."[31]

In fairness to the assistant superintendents, it must be noted that a few took their responsibilities quite seriously but were handicapped by the lack of support they received. One young man stationed in the geyser region asked the superintendent if he might have a piece of equipment that he lacked—a horse. "I would not ask you for a horse if I did not need one," he added apologetically. Congress's appropriation for the assistant superintendents did not extend to their equipment. Those assistants who could afford it sometimes used their own money to procure mounts.[32]

More crippling than lack of experience or equipment was a more fundamental failure in the park. As one assistant superintendent explained, "some lawless men know, or seem to know, that there is no legal authority to arrest and punish for any violation of park regulations." The act creating Yellowstone had bestowed on the Secretary of the Interior the power to create rules, but it gave the secretary no authority to enforce them. Despite the repeated requests for new legislation to repair this gaping hole, Congress failed to take action. As a result, even in the rare instance when poachers were captured, the only action that park officials could take was to seize their contraband (usually skins or heads) and whatever personal property they might have in their possession. Poachers could then be expelled from the park, but most of them, of course, simply re-equipped and came back.[33]

Frustrated with his lack of legal authority to take tough action, Assistant Superintendent Edmund Fish wrote in his annual report that "I am very anxious for power to act like a man and not like a sneak in park matters and hope you can send good news from Congress soon." Another assistant was ready to let someone else try. "Let the Military have charge of the Park."[34] Colonel Ludlow, of course, had proposed the same thing after his 1875 park expedition in the company of Grinnell.[35]

Not surprisingly, senior military officials also supported the idea of a power grab at the expense of the rival Department of the Interior. General Sheridan, having earlier waged a total war against the Indians that contributed to the near extermination of the buffalo, now advocated the animal's protection. "If authorized to do so," wrote Sheridan in an 1882 report, "I will engage to keep out skin hunters and all other hunters . . . and, if necessary, I can keep sufficient troops in the park to accomplish this object."[36]

In the December 14, 1882, edition of *Forest and Stream*, Grinnell publicized both Ludlow's 1875 report and General Sheridan's new call for military control. Sheridan had also called for a major expansion of the park. "We wholly agree with him," wrote Grinnell, "that the limits of the Park should be extended, and that the duty of policing it should be entrusted to the troops under his command. It seems impossible to hope for anything in the way of assistance from the Interior Department, that sink of corruption which so disgraces our government."[37]

Less than a month after Grinnell published the editorial, Senator George Vest took up the call, advocating "one or two companies of cavalry" to dog the "mercenary wretches" killing game in the park. Just as significant, Vest introduced a bill to create explicit police jurisdiction in park territory, including provisions for the prosecution and punishment of poachers. A companion bill was introduced in the House, but both bills died without ceremony.[38]

An 1883 bill resulted in a far more modest provision, authorizing the Secretary of the Interior to *request assistance* from the Secretary of War "to prevent trespassers or intruders from entering the park for the purpose of destroying the game or objects of curiosity therein."[39] Given the rivalry between these departments at the time, of course, the Secretary of the Interior would never make such a request of his own volition. Future circumstances, however, would make this provision critically important.

In the wake of Senator Vest's failure to pass meaningful enforcement measures, Grinnell lamented Congress's "strange apathy in regard to a matter of such vital importance as the preservation to this country of the Yellowstone National Park." Though the editor of *Forest and Stream* worked in New York, he was learning a great deal about Washington. "It is very evident, as it has been from the first, that the monopolists have a strong lobby." The railroads, mining interests, and concessionaires who opposed park protection included what Grinnell admitted were "some of the sharpest intellects, some of the best business ability in the country." Moreover, they were backed up by "unlimited money with which to influence legislation; they had an enormous political power behind them."[40]

THERE IS NO RELIABLE COUNT OF THE BUFFALO THAT STILL ROAMED Yellowstone Park in the mid-1880s. The limited patrolling by park officials made it impossible to obtain an accurate number, but an estimate of 1,000 is probably close to the mark.

However foggy the precise figure, the failure to establish meaningful enforcement in Yellowstone would make the future of these animals crystal clear. "The Park is overrun by skin hunters, who slaughter the game for the hides," wrote Grinnell in *Forest and Stream*, "and laugh defiance at the government." Yellowstone, "in thorough policing by good men, could be made a permanent breeding place for the larger wild animals which will otherwise, before long, become extinct."[41]

In the battle to save Yellowstone Park and its resources—including the last buffalo in the American wild—Grinnell was not without weapons of his own. *Forest and Stream*, under Grinnell's editorial leadership, was emerging as one of the most influential publications in the country. And in Senator Vest, Grinnell had a well-placed and energetic ally.

But it was not enough, and the defeat of Vest's enforcement bill provided Grinnell with a painful demonstration. *Forest and Stream* was helping Grinnell to coalesce a constituency for wildlife, but what that constituency needed was a lobby of its own.

CHAPTER ELEVEN

"The Meanest Work I Ever Did"

—GEORGE BIRD GRINNELL,
ON LOBBYING

I n 1883, the year that the northern herd was extirpated on the plains of Montana, William F. Cody launched the traveling extravaganza that would come to be known as *Buffalo Bill's Wild West*. Featuring wagon trains, simulated Indian fights, trick shooting, and fancy riding, the Wild West show was everything Americans believed the West to be—wanted the West to be.

Bill Cody had been famous since long before 1883—dime novels and an East Coast acting career had seen to that. The Wild West show, though, would catapult the former buffalo hunter to unimagined heights, not just in the United States but also around the world. When Buffalo Bill toured France, the promotional posters showed his face, superimposed on a charging buffalo. Only two words appeared on the poster: *"Je viens"* ("I'm coming"). It was all he needed to say.

Like today's Hollywood stars, Buffalo Bill occasionally used his fame to promote a political cause, understanding well that his pronouncements would find a wide audience. In 1883, Bill Cody wrote a letter to the editor of the *New York Sun*, attached his name to the cause of wildlife protection in Yellowstone National Park. The indiscriminate killing of game, wrote the man who once shot sixty-nine buffalo for the amusement of champagne-sipping spectators, "does not find favor in the West as it did a decade or so ago."[1] Clearly, popular attitudes were beginning to change.

Second thoughts about buffalo hunting by such popular figures as
Bill Cody helped to change public sentiment, but it was not enough to
overcome the power of entrenched Washington lobbyists.

The destruction of virtually every buffalo in America had shown the inexhaustibility of natural resources for the myth that it was. While Americans in the 1880s still celebrated the achievement of settling the West, other sentiments had begun to creep into the national consciousness. Grinnell's *Forest and Stream* editorials, for example, would frequently convey a tone of nostalgia and sadness. "The rapidity with which the Rocky Mountain region is being settled up is astonishing to anyone who has for years been familiar with it, and who remembers a time when it was a simple wilderness, whose only tenants were the wild beasts or still wilder men."[2]

Having tamed the West, many Americans now found themselves wishing they had left it a bit more wild. No shock fell more profoundly than the destruction of the buffalo, a species whose multitudes so recently had seemed unimpeachable. By the mid-1880s, Americans knew better. But nostalgia was not the same as political power, and on the issue of protecting wildlife, Grinnell and his Washington allies could muster no better than a draw.

The legislative stalemate that began in 1883 would repeat itself for a decade. On Grinnell's side were advocates of new enforcement measures—their central cause a law to create tough penalties for the destruction of game in Yellowstone National Park. On four occasions between 1883 and 1890 the U.S. Senate actually passed a bill that would have achieved this goal. Two of these bills passed the Senate unanimously. All four bills, however, were blocked in the House of Representatives by the railroad lobby. Each time its tactic was the same: Attach to the Senate's enforcement legislation an amendment granting a railroad right-of-way through the park. No right-of-way, no enforcement bill. Supporters of enforcement refused to accept the deal, unwilling to sacrifice a portion of the park as the price of winning new protection. So year after year, the deadlock continued.[3]

A FREQUENTLY REPEATED WASHINGTON MYTH ATTRIBUTES THE VERY term *lobbyist* to the railroad wheeler-dealers of Gilded Age America. According to lore, the term was invented by President Grant, who frequently strolled from the White House to the nearby Willard Hotel for lunch. Whenever arriving at the Willard, Grant was descended on by a flock of railroad agents, waiting to pounce from their perches in the lobby. Though Grant may well have employed the term *lobbyist*, it appears already to have been in use decades earlier to describe the lobby-residing favor-seekers outside of Britain's parliament chambers. Regardless of whether the Willard Hotel birthed the term, the story about President Grant paints a telling picture of the power and influence of railroad lobbyists in the Gilded Age. Modern American lobbying was invented in this era, and the railroad sector played the leading role.

The prize, for railroads, was straightforward—the massive public financing necessary to launch projects of epic scale. To win this prize, the railroads deployed veritable armies of lobbyists at the nation's capitol. Lobbyists, in the Gilded Age, were dubbed the Third House of Congress. The Texas and Pacific Railway—a modest company compared with the likes of such giants as the Northern Pacific—kept 200 lobbyists on its books.[4]

While the purchase of influence through naked bribery (of the cash-in-an-envelope variety) was not unknown in the freewheeling Congress of the 1870s and 1880s, more common was an arsenal of subtler but equally corrupting measures.

Rules against conflict of interest did not exist in this era, and many congressmen were also businessmen with a direct financial stake in the outcome of measures on which they voted. Thomas Bayard, a senator from Delaware, complained in 1882 that a senator could "vote upon a question involving his direct personal interest" or "bring into Congress schemes and plans which should put money into his own pocket."[5]

If a member of Congress did not have his own financial stake in railroad ventures, the railroads often created one for him through gifts of stock. This was a particularly attractive form of bribery for the railroads, notoriously stock-rich but cash-poor. The most common transfer involved selling shares to members at below-market value. At a time, for example, when a share of the West Virginia and Pittsburgh Railroad company sold for $40 on the open market, the company offered shares to senators for $20.[6] Senators could sell the stock for an immediate windfall, and those who held the stock inherited an ongoing financial interest in the company's success.

In addition to stock, congressmen who voted with the railroads might receive a variety of other financial rewards. No rules forbade lobbyists from lending money to members of Congress, and such loans were common. Members whose careers marked them as friends of the railroads might also expect to be "taken care of" when they ended their term in office. In reference to a retiring senator, a Central Pacific lobbyist wrote headquarters of the need to "arrange something out of which he could make some money (something handsome)." Other Central Pacific allies were given land, another asset that the railroads held in large quantity. Central Pacific officials gave Senator William Stewart of Nevada property in one of their railroad towns. Senator Aaron Sargent of California received "fifty thousand acres of land of average quality of the lands along the line of the road."[7]

In 1886, a bill was introduced into the Senate that would have forbidden "employment or payment for services from a railroad company or any of its officers which had obtained its charter or grant of land or pecuniary aid from the United States Government." It was voted down.[8]

In 1887, another bill to clamp down on railroad abuses succeeded—

at least partially. Like today, congressmen of the nineteenth century enjoyed the perquisite of subsidized travel. Railroads dolled out free passes by the bushel basket, so many that it began to hurt their profits (a fact that probably helped to explain the passage of this particular reform). A Central Pacific official complained about the burden of having carried 6,186 "deadheads." The new law, however, included a loophole for members of Congress who were also "railroad officials." Not surprisingly, a number of congressmen soon acquired titles. Senator Richard Pettigrew of South Dakota, for example, was anointed "President of the Sioux Falls Terminal Railroad"—a ten-mile stretch of rails. Thus he continued to travel on the Northern Pacific dime.[9]

In this wide-open era, according to historian David Rothman, Congress at times "resembled the floor of a brokerage house."[10] It was in this domain that the fate of legal protection for America's last wild buffalo would be decided.

GEORGE BIRD GRINNELL HAD A HEALTHY DISDAIN FOR LOBBYING. It was "the meanest work I ever did," he wrote in a letter to his friend Lute North. "I would rather break bronchos for a living than talk to Congressmen about a bill. It makes me feel like a detested pickpocket to do it."[11] Like it or not, politics—including the tawdry world of lobbying—would prove essential to Grinnell's high-minded goals.

Grinnell brought great strategic skills to the political battlefield. As any good politician knows, prevailing in Washington often turns on the threshold action of framing the debate. Grinnell framed the debate over the commercialization of Yellowstone Park, for example, as a contest between "monopolistic land-grabbing schemers" and "the interest of the people." In discussing game protection laws, Grinnell shifted the focus from what was being taken away—the unfettered right to kill game—and placed it instead on what was being conserved—the ownership of every American citizen in federal resources. Grinnell also showed political savvy by placing the issue of game protection in the unassailable context of conserving for future generations, or, as he put it, the "grandchildren of the present generation."[12]

As a political strategist, Grinnell also was skilled at framing issues in consonance with the prevailing orthodoxy of his times. Historian H.

Duane Hampton described the Gilded Age as an era in which "exploitation of the nation's resources was a way of life . . . Americans worshiped the practical and the useful, and generally held altruism and aestheticism in disdain."[13]

Grinnell's writings show his own near-religious appreciation for the aesthetic beauty of wild places, but he knew that in robber-baron America, this view carried little political weight. In a mid-1880s battle over protection of the Adirondacks, for example, Grinnell argued that the "reasons why the forests should be preserved are not sentimental, but very practical." As illustrations, he emphasized the necessity of forests in protecting the water supply for New York City. By similar token, Grinnell maintained that one important reason for preserving forests from unregulated lumbering was the necessity of ensuring a steady supply of building materials for the future.[14]

Grinnell and other early conservationists are sometimes criticized as narrow-minded utilitarians, insufficiently pure in their advocacy for the environment. The reality is that Grinnell understood that pragmatism was essential to getting things done.[15] Grinnell was not interested in convincing every American to see the environment in the same way he did. What he cared about were results.

A significant part of achieving results, as Grinnell understood instinctively, involved forging strategic alliances—what political strategists today call coalition-building. In one of Grinnell's earliest efforts to promote game protection, he recognized the importance of an alliance with farmers. One of his editorials carried the headline "The Mutual Interests of Farmers and Sportsmen." He wrote, "Let all the farmers and their sons and hired men find it pays them to protect birds and they will do it and satisfy all parties."[16]

In the West, Grinnell framed the issue of game protection in utilitarian terms that appealed to a powerful group of potential allies—ranchers. "A sentiment against the unnecessary killing of game is growing up in the West, especially among stockmen," Grinnell reported. "These men are beginning to realize that while they can supply their ranches and camps with wild meat they are putting a certain amount of absolute cash in their pockets; that every time they bring in an elk, it saves them a steer." Indeed, groups such as the Wyoming Stock Growers' Association were sometimes important allies in the battle for game protection laws and enforcement.[17]

The economic benefits of game protection accrued not only to ranchers, argued Grinnell, but also to entire communities. In an editorial titled "Dollars and Cents," Grinnell described the wide range of businesses to benefit from sportsmen-tourists, including railroads, guides, hotels, and outfitters. (Today, of course, a whole industry has grown up around ecotourism.) Without game protection, warned Grinnell, the economic benefits of this trade would "go to Canada."[18]

On the front page of the May 15, 1884, edition of *Forest and Stream*, Grinnell revealed two insights that foreshadowed much of his political work in the decade to follow. One observation arose in the context of New York State affairs, though he would soon apply it to the national stage. Grinnell argued that "there ought to be a live association of men bound together by their interest in game and fish, to take active charge of all matters pertaining to the enactment and carrying out of the laws on the subject."

Grinnell's second insight showed his keen appreciation of his fellow Americans' social tendencies. "Americans delight in national associations," he wrote. "Almost every branch of trade has its organization of this kind, and so has almost every branch of 'sport.'"[19]

In 1886, Grinnell announced the creation of a national association that would stand as one of his most lasting legacies. Big game animals such as buffalo were not were not the only wildlife under pressure in nineteenth-century America. As Grinnell frequently reminded his readers, one of the most common targets of commercial hunters was birds. In Chesapeake Bay and other flyways, hunters used canon-sized shotguns to mow down entire flocks of ducks and geese as they floated on the water. "Pigeoners" used nets to kill passenger pigeons in numbers sufficient to fill railcars, a practice that would lead to that species' extinction. And songbirds of all variety fell victim to hunters who killed to feed a fashion craze that began in the 1870s—women's hats adorned with colorful feathers (and even whole birds).[20]

"The land which produced the painter-naturalist, John James Audubon, will not willingly see the beautiful forms he loved so well exterminated," wrote Grinnell in the February 11, 1886, edition of *Forest and Stream*. "We propose the formation of an association for the protection of wild birds and their eggs, which shall be called the Audubon Society." Grinnell encouraged interested persons to "establish local societies for work in their own neighborhood." The idea was an immediate success. In February 1887, Grinnell launched *Audubon Magazine*, whose first issue announced 20,000 Audubon Society members in 400 communities nationwide. By 1888, membership had grown to 50,000.[21]

In launching the Audubon Society, Grinnell catalyzed a new cadre of conservationists. While critical to his continuing efforts to create a constituency for wildlife, however, the early Audubon Societies were not explicitly political. Grinnell still had no lobby, a fact that he, together with an up-and-coming New York politician, were about to change.

IN THE JULY 2, 1885, EDITION OF *FOREST AND STREAM*, GRINNELL reviewed a book for a popular section called "New Publications." As Grinnell would later remember, "it was the writing of a review of *Hunting Trips of a Ranchman* that brought me into intimate contact with Theodore Roosevelt."[22] *Hunting Trips of a Ranchman* was a memoir detailing the adventures of twenty-six-year-old Roosevelt, the New York State assemblyman who had retreated to ranch life in the Dakotas following the tragic deaths of both his wife and his mother in the course of a single day.

Grinnell's review took note of Roosevelt's "excellent work" in Albany, where he was known as "an earnest and energetic politician of the best type." Grinnell, a grizzled thirty-five, forecast a bright future for young Roosevelt. He noted that Roosevelt's work in Albany had shown him to be "a person of exceptionally well-balanced mind, and calm, deliberate judgment, and these qualities cannot fail to make their impression in any pursuit."

If Roosevelt was annoyed by the patronizing tone, he was about to become more so when Grinnell turned to a critique of the book itself. "Mr. Roosevelt is not well known as a sportsman, and his experience of the Western country is quite limited, but this very fact in one way lends an added charm to his book," wrote Grinnell. "He has not become accustomed to all the various sights and sounds of the plains and the mountains, and for him all the difference which exists between the East and the West are still sharply defined." In other words, Theodore Roosevelt was a greenhorn.

A young Theodore Roosevelt
in full frontier regalia,
as he appeared around the
time he wrote *Hunting Trips
of a Ranchman.*
*Courtesy of the National
Park Service, Yellowstone
National Park.*

"We are sorry to see that a number of hunting myths are given as fact," continued the review, "but it was after all scarcely to be expected that with the author's limited experience he could sift the wheat from the chaff and distinguish the true from the false."[23]

Roosevelt, hardly a man to let a punch go unanswered, stormed into *Forest and Stream*'s offices to confront Grinnell. The two men, according to Grinnell's recollections, "talked freely about the book, and took up at length some of its statements." The details of their conversation are not recorded, but Grinnell would later explain that what Roosevelt saw "he reported faithfully and accurately, but he had the youthful—and common—tendency to generalize from his own observations and to conclude that certain aspects of nature were always and in all places as he had found them in one place." Roosevelt, at least by Grinnell's account, "at once saw my point of view."[24]

After their lively deconstruction of Roosevelt's book, Grinnell remembered how the two men moved to a broader discussion of their mutual passion for hunting in the West. They also discussed game protection. "I told him something about game destruction in Montana for the hides, which, so far as small game was concerned, had begun in the West only a few years before that, though the slaughter of the buffalo for their skins had been going on much longer and by this time . . . the last of the big herds had disappeared."[25]

Grinnell and Roosevelt had much in common. Their upbringings were similar—both raised in wealthy New York families and both interested, from an early age, in the outdoors. Roosevelt, remembered Grinnell, "was always fond of natural history, having begun, as so many boys have done, with birds." Like Grinnell, Roosevelt had also been influenced by his boyhood attachment to the pulp novels of Mayne Reid.[26]

In the months and years that followed, Roosevelt began to call frequently on Grinnell at *Forest and Stream* headquarters, where increasingly their discussion focused on the destruction of game in the West. The two men agreed that big game animals were in jeopardy "as soon as a point should be reached where their products could be turned into dollars." Grinnell shared a copy of the 1875 report he had written about his expedition to Yellowstone National Park, including its early call for game protection. Grinnell remembered how this "much impressed Roosevelt, and gave him his first direct and detailed information about this slaughter of elk, deer, antelope, and mountain sheep." Indeed Grinnell's influence on Roosevelt would have far-reaching implications for the broader history of conservation.[27]

Grinnell would later write that "those who were concerned to protect native life were still uncertainly trying to find out what they could most effectively do, how they could do it, and what dangers it was necessary to fight first." As the months passed, and particularly after the success of the Audubon Society in 1886, Roosevelt and Grinnell gravitated toward the formation of a national organization centered around sportsmen.[28]

On a December evening in 1887, Theodore Roosevelt hosted a dinner party in his Manhattan home for a prominent group of hunters, among them his friend Grinnell. Roosevelt proposed "the formation of a club of American hunting riflemen, to be called the Boone and Crockett Club." In January 1888, the organization adopted a constitution and elected Roosevelt as its president. February 29, 1888, was a seminal moment in the history of conservation: the first official meeting of the first-ever organization to be formed with the explicit purpose of affecting national legislation on the environment.[29]

Some of the club's objectives sounded a lot like Roosevelt: "To promote manly sport with the rifle." Some sounded a lot like Grinnell: "To work for the preservation of the large game of this country, and so far as possible to further legislation for that purpose, and to assist in enforcing the existing laws."[30]

The minutes of the first Boone and Crockett meeting underscored both the club's explicitly political aim and the attention that it would focus on Yellowstone. The members passed a resolution naming Grinnell and four others to a committee whose goal was "to promote useful and proper legislation towards the enlargement and better government of the Yellowstone National Park."[31] It was the only specific initiative accorded such attention.

To achieve their political goals, it was clear, Boone and Crockett would play insider pool. Grinnell, Roosevelt, and club secretary Archibald Rogers were elected as the "committee on admission," and a decision was later made to limit the club to 100 members. Grinnell's announcement of the club in *Forest and Stream* was revealing: "It would seem that an organization of this description, composed of men of intelligence and education, might wield a great influence for good in matters relating to game protection." Among Boone and Crockett's first members were General William Tecumseh Sherman, Senator George Vest, and Edward F. Beale—a prominent Washingtonian who owned the Decatur House, a half block from the White House. These were men, according to Grinnell, whose "names carried weight."[32]

In addition to Grinnell's formal responsibilities in Boone and Crockett, he also acted as the unofficial spokesman-publicist, using *Forest and Stream* as "the natural mouthpiece of the club."[33]

With the formation of Boone and Crockett, Grinnell hoped, all the pieces were in place.

While not exactly on equal footing with the railroads, Grinnell now had an organized lobby of heavy-hitting insiders to carry the case of conservation.

Year by year, Grinnell and his allies were gaining traction against the prevailing robber-baron orthodoxy. "Those who used to boast of their slaughter are now ashamed of it," wrote Grinnell in 1889, "and it is becoming recognized fact that a man who wastefully destroys big game, whether for the market, or only for heads, has nothing of the true sportsman about him."[34]

But the question still open was whether it all came too late. Grinnell now had his lobby, but the congressional stalemate over enforcement legislation for Yellowstone had not been broken. And the battle in Washington, or, more accurately, the war of attrition, was precisely the type of contest that the buffalo, as a simple matter of numbers, could not sustain.

WHILE WASHINGTON DEADLOCKED, THE GOVERNORS OF MONTANA and Wyoming—the two territories with the greatest interest in Yellowstone—had become increasingly aggravated with the ambiguous legal status of the national park. Part of the frustration stemmed from failed aspirations on the part of both territories to acquire sovereignty over the park (the two territories were rivals in this regard), but part also grew out of disgust with the failure to stop the widespread poaching of park resources.

Against this backdrop, a Yellowstone St. Patrick's Day party set in motion a chain of events that though ending in sheer farce, gave the Park its first, brief taste of meaningful enforcement. It all started when a drunk named David Kennedy shot a man named James Armstrong near park headquarters. The Yellowstone superintendent informed the governor of Wyoming, William Hale, of the shooting. By the time Wyoming received the news, David Kennedy was nowhere to be found. Finally at wit's end over Yellowstone's mishmash of legal authority, Governor Hale asked the next session of the Wyoming territorial legislature to take action, and they obliged him.[35]

A statute of March 6, 1884, extended the jurisdiction of Wyoming law over Yellowstone National Park. One aim of the new law, according to the territorial legislature, was "to protect and preserve the timber, game, fish, and natural objects and curiosities of the park." It also provided for two constables and two peace officers to oversee park territory—and even appropriated money for their salaries and equipment. Most important, the Wyoming legislature did what the U.S. Congress had not: established penalties for poachers, including six months in jail.[36]

Yellowstone's formerly impotent federal officials welcomed the new law with cheers. Josiah W. Weimer, one of the much-maligned assistant superintendents, crowed that the Wyoming law "put a club in my hands." The statute quickly resulted in the first prosecution for poaching in the history of Yellowstone—twenty-two years after the park's creation. A tough-bitten scout named Edward Wilson caught a party of notorious poachers in the possession of elk meat and beaver skins and promptly hauled them before one of the newly appointed justices of the peace. The justice of the peace issued fines and sentenced one of the men to the full six-month jail term allowed by the law.[37]

Even some Washington officials applauded the law. After seeing it in action, a group of touring congressmen returned to the Capitol and advocated that the law be given formal federal blessing. In fact, they went so far as to advocate turning the park over to Wyoming when the territory became a state. Yellowstone's superintendent at the time, David Wear, used his annual report to complain about the lack of federal enforcement measures. "I would most earnestly call your attention to the entire inadequacy of the laws to provide punishment for violations of the regulations for the protection of the Park," he wrote. "The protection that I have been able to give the Park has been through the Territorial laws of Wyoming."[38]

Despite these endorsements of the Wyoming law, there were also a few flies in the ointment. Frontier justice remained a tad crude, and Wyoming's territorial legislators had included a few provisions in their statute that fared poorly under the bright light of eastern pettifoggers. One section, for example, gave the justices of the peace the right to keep all fees collected when adjudicating under the statute. Another allowed witnesses for the prosecution to keep half of any fines assessed if a defendant were found guilty.[39] These financial incentives in favor of conviction struck some as a bit injudicious.

The whole matter came to a head in 1885, when eastern VIPs ran afoul of the new Wyoming law. A party of eleven Yellowstone campers— including Lewis E. Payson, a congressman from Illinois and a former judge, and Joseph Medill, editor of the *Chicago Tribune* and a former mayor—was charged with leaving a campfire without properly extinguishing it. The man pointing the finger of accusation, a constable-schemer named Joe Keeney, had been making a nice living by collecting on the proceeds of zealous enforcement.

The case of Congressman Lewis E. Payson and party was heard in the chambers of a justice of the peace named Hall. (The relationship of Justice Hall to Wyoming's Governor Hale is unclear, but the experience of the justice of the peace prior to his judicial appointment had been as a woodcutter.) Despite considerable testimony that the Payson party fire had been properly doused, Justice Hall found the men guilty. After consulting with Constable Keeney about the appropriate fine, the justice assessed a penalty of $60 (half of which would go to Keeney) and court costs of $12.80.[40]

Congressman Payson, condemning the "kangaroo court," promptly announced his intention to appeal and offered to post a $1,000 bond. This development threw the whole proceeding into considerable disarray. A stunned Justice Hall, who had never encountered a defendant quite like Congressman Payson, actually turned to Payson for legal advice. At this point, editor Medill launched his own verbal jab at Justice Hall, calling him a "damned old Dogberry." The reference, to the incompetent judge in Shakespeare's *Much Ado About Nothing,* flew over Justice Hall's head. Still, the justice was smart enough to suspect an insult and called for a dictionary.

By this time, though, Justice Hall's top priority was to clear Congressman Payson's case off his docket. He offered to reduce the fine to $10, a plea bargain that Payson refused. The justice next offered to accept a fine of $1 along with any amount for court costs that Payson considered fair. Payson steadfastly refused to pay even a dollar in fine but did offer up a few bucks for court costs.

The episode received considerable play in the newspapers. Editor Medill ensured that the affair was publicized in the *Chicago Tribune,* and local papers too got a good laugh at the expense of poor Justice Hall. The editor of the *Livingston Enterprise* explored the Shakespearean origins of Medill's insult, opining that the "Dogberry" epithet provided "a very faithful mirror of the administration of justice at Norris Basin."[41]

The whole sorry event might have been filed away as an amusing anecdote, except the publicity and the chain of events it helped to spark resulted in a devastating consequence. A report commissioned by the

Secretary of the Interior, with the *Chicago Tribune* article on Justice Hall appended, concluded that the territory of Wyoming had no authority to enforce its laws within the boundaries of Yellowstone. On March 10, 1886, the Wyoming legislature repealed its Yellowstone statute. The brief experiment with tough enforcement had come to an end.

The widespread attention to the flow and ebb of enforcement was like ringing the dinner bell for poachers. In the view of Yellowstone historian Aubrey Haines, the repeal of the Wyoming law "left the area in a worse plight than it had ever been." Poachers flooded back into the park, now even more aware that they faced no meaningful consequence if caught. In addition to the upsurge in poaching, there was another dire new development. A few of the most hardcore miscreants around the park deliberately set grass and forest fires with the vengeful goal of embarrassing Yellowstone's civilian leadership.[42]

By the summer of 1886, the anarchy in the nation's first national park had reached such proportions that even Washington took notice. After fourteen years of civilian administration of Yellowstone by a Department of the Interior that ranged between feckless and corrupt, there was about to be a new sheriff in town. The cavalry was coming.

CHAPTER TWELVE

"A Terror to Evil-Doers"

*He has ruthlessly turned out of the Park the bad characters
that at one time infested it, and has made his command a
terror to evil-doers.*

—GEORGE BIRD GRINNELL ON
CAPTAIN MOSES HARRIS[1]

O
n August 17, 1886, Captain Moses Harris and the fifty members of his command—Company M, First U.S. Cavalry—rode
tired mounts into Yellowstone National Park. By evening the
cavalrymen reached Mammoth, pitching their tents at the foot of the
famous hot springs. The brimstone steam and the chalky-white, terraced
hillside were not the only sights to captivate the attention of the newly
arrived soldiers. They could also see the fiery glow of a burning mountain, less than seven miles away, one of the "vengeance" fires set by the
poachers who ringed the park. Three days later, the civilian superintendent of the park, David Wear, relinquished his authority to Captain Harris. For the next three decades, the U.S. Army would preside over the
nation's first national park, and on the cavalry would rest the tenuous
hopes for saving the buffalo.[2]

The story of how the U.S. Army came to occupy Yellowstone can be
traced, in no small part, to the misadventures of Congressman Lewis E.
Payson in the kangaroo court that was Yellowstone's justice system. The
Illinois congressman returned to Washington with a capitol-sized chip
on his shoulder. And in Washington, Payson was far from alone in his

hostility toward the park. Payson led the so-called railroad gang and was, in the description of George Bird Grinnell, a "prominent member of the Public Lands Committee." Payson and other opponents of the park found a focal point for their anti-Yellowstone animus in the form of the 1886–1887 park appropriation bill.[3]

In the House of Representative's consideration of the bill, park opponents succeeded in cutting the Yellowstone appropriation in half, from $40,000 to $20,000. Even more worrisome, congressional debate of the bill demonstrated that more than two decades after the creation of the park, its continued existence was far from secure. In both the House and the Senate, members argued on the floor for Yellowstone's outright abolition. Senator Preston Plumb of Kansas, for example, advocated that the whole park be "lopped off," the lands to be made available for mining, homesteading, or other private purposes. Representative John Reagan of Texas concurred, arguing that it was not the role of government to engage in "show business."[4]

The debate over the future of the park eventually wound its way into a heated conference committee, where supporters of Yellowstone, led by Grinnell's ally, Senator George Vest of Missouri, succeeded in preventing the park's wholesale dismemberment. As for funding, though, the appropriation would stay at $20,000, with an additional, onerous proviso: Monies could be used only for the construction of roads, meaning that no funds were available for the payment of the park's civilian administrators.

Under the 1883 statute, Congress had given the Secretary of the Interior the right to request troops from the Department of War for the purpose of upholding law in the park. Given the rivalry of these departments, no such request had ever been forthcoming. Now, though, Congress's elimination of funding for civilian administrators forced the hand of the Secretary of the Interior, L.Q.C. Lamar. Lamar penned the appropriate letter to the Secretary of War, and a few weeks later, Captain Harris and his Company M "boys in blue" were bound for Yellowstone.[5]

New sheriff in town: Captain Moses Harris leads Company M,
First U.S. Cavalry, into new territory—Yellowstone National Park . . .
and a new mission—the battle against poachers.
Courtesy of the National Park Service, Yellowstone National Park.

Moses Harris, the first military superintendent of Yellowstone
National Park, was a native of New Hampshire and a veteran of the
Civil War. For his actions in a battle at Smithfield, Virginia, near the end
of the war, Harris was awarded the Congressional Medal of Honor.
According to the official citation, "In an attack upon a largely superior
force, his personal gallantry was so conspicuous as to inspire the men to
extraordinary efforts, resulting in complete rout of the enemy."[6]

Though a combat veteran and war hero, little in Harris's military
career had prepared him for the type of duties he faced in Yellowstone.
Yet he set about his new mission with the energy and zeal that defined
him. Having been placed in charge, Captain Harris did what all good
leaders do—he took charge. In his triage of the hemorrhaging park, the
captain's first priority was to fill the vacuum created by Congress's attack
on funding. News that the park's civilian employees would not be paid
had traveled quickly, deepening the anarchy that already existed. In the
days before Harris's arrival, his civilian predecessor, David Wear, dispatched a panicked telegram to the Secretary of the Interior reporting
that "lawlessness in Park has rapidly increased." Not surprisingly, many
of Wear's assistant superintendents abandoned their posts after learning
that they would not be paid.[7]

Accompanied by Wear, Captain Harris conducted a quick tour of the
park, depositing soldiers at key tourist sites, including Soda Butte (near
Cooke City), the canyon falls, and the geyser basins. Harris's other
immediate priority was to battle three forest fires then raging in Yellowstone. Captain Harris agreed with Wear that the fires had probably been
set by "unscrupulous hunters." In addition to revenge, the captain
believed that one motive for the poachers' fires was more practical—to
drive game outside of the protected domain of the park.[8]

Back from his quick reconnaissance of the park, Harris issued Order
No. 5, a set of regulations that would form the foundation for more formal rules issued by the Department of the Interior in 1887. Order No. 5
forbade "hunting or trapping and the discharge of firearms within the
limits of the Park." The sale of fish from the park was prohibited, and fishing allowed only "with hook and line." Campfires were to be built only

"when actually necessary" (a rather amorphous restriction). Livestock was to be kept away from "points of interest." Finally, the order attempted to quell the popular tourist activity of dumping debris into geysers in order to test their strength: "No rocks, sticks, or other obstructions must be thrown into any of the springs or geysers within the Park."

Perhaps as significant as the individual rules (some of which had been in place with earlier superintendents) was the manner in which Captain Harris ordered his men to conduct themselves. "They will in the enforcement of their orders conduct themselves in a courteous and polite, but firm and decided, manner. They will not hesitate to make arrests when necessary, reporting at once ... to the Commanding Officer."[9] In short, Harris aimed to replace the haphazard incompetence (and worse) of the civilian era with a tough new military professionalism.

Having arrived in August, Company M spent the first two months of its Yellowstone tour focused on the twin occupations of managing tourists and fighting forest fires. By early fall, though, deep snow brought both the tourist season and the fires to an end. Captain Harris, as he reported to Washington, shifted his men from the geyser basins "for the important duty of affording protection to the large game which was being driven from the mountains by the early and unusually heavy snowfall."[10]

Snowfall, to Yellowstone poachers, was the signal for open season. With winter cold, hides thickened into plush robes that were most valued by the market. Snowfall also drove game from high-mountain, inaccessible summer habitat into the broad river valleys. Deep snowfall—ten feet was not uncommon—made it difficult for heavy animals such as buffalo to flee. Hunters could approach to within a body length of floundering animals before easily dispatching them. Traditionally, winter had also been a season in which park officials did not conduct patrols, retreating perforce into the sheltered enclave of Mammoth. As poachers fully realized, the chance of capture during winter was virtually null.

Captain Harris was determined to alter this calculus. One of Harris's earliest actions was to hire a seasoned scout. The captain had wanted three—hardly unreasonable given Yellowstone's size—but was allowed only one. In *Forest and Stream*, Grinnell emphasized the particular importance of experienced guides for new troops. "The new troops, who are ignorant of the region will find themselves unable to cope with mountaineers [i.e., poachers] familiar with almost every foot of the ground in the Park, and unless the different scouting parties are furnished with efficient guides, much destruction of forests and game may result."[11]

Though limited to one scout, Captain Harris at least succeeded in hiring a top-notch man, Edward Wilson. Wilson was himself a former trapper, and knew well both the habits and the hideouts of the men he now hunted. In his official reports, Harris repeatedly praised Wilson for his "faithful, zealous, and courageous performance of the difficult duties with which he has been charged."[12]

Poachers, in Captain Harris's description, "surround the Park." Many hunters established semipermanent camps just outside of the park boundaries, waiting for game to wander out—or crossing into protected territory when the opportunity presented itself. In his 1887 report, the captain described the "great activity and vigilance by scouting parties" required "to prevent [poachers] from operating within the borders of the Park." To the degree that he could patrol the border regions in harsh weather conditions, Harris dispatched his cavalrymen to make preemptive contact with hunters' camps and to "warn them off." Harris believed that this practice was successful, reporting that "it is believed that little or no game was killed within the Park."[13]

While the cavalry's winter patrols were a welcome step, it appears virtually certain that many poachers were slipping through. Hunters were often apprehended in the park with meat or hides—usually claiming that the game had been taken outside Yellowstone's borders. Unless the hunters were caught in the act (an exceedingly rare occurrence), the law left the soldiers no alternative but to let their quarry go.[14]

Poachers also took advantage of a vital tool of winter transport that the army had yet to utilize in the park. The poacher's winter vehicle of choice was cross-country skis. In the language of the nineteenth century, such skis were called snowshoes, Norwegian snowshoes, or Norwegian skis. (What we think of today as snowshoes were then called webshoes and were and considered a far inferior alternative.) Skis of the day consisted of ten-foot wooden slabs, tapered at the tip. A strap held the toe near the middle of the ski, and the bottom was coated with melted wax or grease to improve gliding. To allow skiers to gain purchase on steep uphill runs, animal skins were sometimes tied to the bottom of the ski, fur side down, with the grain of the fur running front to back. When the ski was moving forward, the grain allowed the ski to slide, but when the ski was moving backward, the hair caught against the snow to provide traction. (The modern, synthetic counterpart is still called a skin.)[15] To propel himself, the skier used a single pole. Wearing skis, poachers could approach to within a few feet of their quarry. No longer did buffalo hunting require a fine rifle.

In its early years of Yellowstone duty, the army struggled to play catch-up with the experienced frontiersmen who made their living by poaching. The hazards of winter patrolling, even for experienced men, were demonstrated in a small 1887 expedition mounted by famous Arctic explorer Frederick Schwatka, who had published several popular books about his adventures. Schwatka led a party of eight men, including a photographer named F. Jay Haynes.

As Schwatka's parties had done in the Arctic, his Yellowstone expedition used skis, pulling their supplies behind them on toboggans. After a sunny start out of Mammoth on January 2, 1887, conditions rapidly deteriorated. On their second night, camping at Indian Creek, the temperature plummeted to thirty-seven below zero. The second day the Schwatka party covered only four miles, and by the third day they abandoned their toboggans. The bitter conditions of these first three days injured Schwatka's lungs, a fact that manifested itself when he began "to hemorrhage profusely." The Arctic explorer was forced to abandon the expedition. The photographer, Haynes, took over leadership of the group, and army scout Edward Wilson was also sent out to assist them.[16]

In his annual report, Captain Harris recorded the success of the Haynes party "in surmounting all obstacles," producing the first winter photographs of the park. As for skiing, Harris concluded that "[t]he difficulties of snow-shoe travel in the Park are such, however, that it is not to be recommended as a winter diversion."[17]

Despite these difficulties, Captain Harris ordered his men to begin learning to use skis, and by the spring of 1887, Company M had something to show for its efforts. In April, the captain learned that a teamster named William James had set up a winter camp in the Norris Geyser Basin. At the end of the fall freighting season, James had traveled to Gar-

The Haynes expedition: Elk, like buffalo, find it difficult to move in heavy snow, making it easy to approach them on skis.
Courtesy of the National Park Service, Yellowstone National Park.

diner and acquired a rifle and traps—then returned to the park. Harris believed he was trapping illegally, so he sent out a ski expedition led by Sergeant John Swan to find out. The sergeant found beaver and lynx skins cached near James's camp, as well as traps set in the nearby Madison River. The soldiers arrested him, and he admitted to illegal trapping with the assistance of two employees of the Yellowstone Park Association (successor concessionaire to the infamous Yellowstone Park Improvement Company). James was brought before Captain Harris in Mammoth, who promptly expelled him from the park.[18] As events would show, however, James would soon resume his Yellowstone criminal career.

BY THE WINTER OF 1887–1888, CAPTAIN HARRIS WAS READY FOR A more ambitious effort to protect Yellowstone's wildlife. For Harris, no species was more important than the buffalo. In his 1888 report he wrote that "the buffalo or bison have so narrowly escaped extinction, and the number which now find a refuge in this Park is so limited, that they should be protected by every possible method."

Captain Harris and his men had seen few buffalo during their first year in the park, the captain admitting that he was "unable to state their numbers with any approach to accuracy." His lack of contact, though, was a source of concern. "My impression is that they have been heretofore somewhat overestimated." From what information he could gather in the summer, Harris believed that the Yellowstone buffalo had been reduced to three small groups—one in the northeast corner of the park along Slough Creek and the Yellowstone River, one in the vicinity of Specimen Mountain-Pelican Creek, and a third, very small group near the Upper Geyser Basin.[19]

To protect the buffalo, Captain Harris needed to know where they were—particularly in the winter season favored by poachers. "[I]t had occurred to me as extremely desirable, if possible, to secure some accurate information concerning the winter haunts of the buffalo which were known to be in the Park." For the dangerous duty, Harris picked his scout, Edward Wilson, and a sergeant named Charles Schroegler. The two men skied out of Mammoth on February 13, 1888. Each man carried a single blanket and a backpack (having learned from the Schwatka debacle that pulling a sled was impracticable). "The hardships of an expedition of this character," wrote Harris, "can only be realized by those who are acquainted with the winter aspect of the mountain solitudes into which these brave and hardy men ventured."

Wilson and Schroegler returned to Mammoth after ten days in the winter wild, having seen only three buffalo. A second ski expedition, though, was more successful. In April, Edward Wilson again trekked out of Mammoth, this time accompanied by a seasoned adventurer—and *Forest and Stream* correspondent—named Edward Hofer. In the protected valleys between the Yellowstone and Madison Rivers, Wilson and Hofer found a herd of more than 100 buffalo. Summer scouting forays confirmed both the presence and the health of this herd.[20]

In his report of 1888, Captain Harris offered a cautiously optimistic assessment of the buffalo in Yellowstone National Park. "From the numbers seen and from the quantity of 'sign' observed over an extended area, the number of these animals that range in this portion of the park can be estimated at not less than two hundred." It was a hopeful turn, yet as Harris fully realized, it was continuing success of his law-enforcement efforts that would determine the buffalos' fate. "The large number of young calves and yearlings which have been seen leads to the belief that a natural increase is in progress, and that if proper protection is afforded the species will not, as has been feared, become extinct."[21]

Certainly Captain Harris was making the most of the resources at his disposal. His men continued to make arrests, and when poachers were caught with sufficient evidence, Harris took pleasure in expelling them from the park. One of those removed, for example, was a poacher caught by Harris's men near Heart Lake after killing an elk.[22]

George Bird Grinnell—no easy sell—found much to praise in Harris's efforts. In a *Forest and Stream* editorial, Grinnell told his readers how Harris had "ruthlessly turned out of the Park the bad characters that at one time infested it, and has made his command a terror to evil-doers."[23]

Yet Captain Harris found himself floundering against the same shoals that had grounded the best of his civilian predecessors. In Harris's 1889 report—the last of his Yellowstone tour of duty—the frustrated captain echoed a familiar refrain: "The inadequacy of mere rules and regulations, unsupported by any appearance of force or penalties for their infraction, soon become apparent." As Harris further noted, "there has been hardly a report rendered relating to the Park during the eighteen years of its existence in which the necessity of some further provision of law for its preservation and government has not been urged."[24]

With enforcement legislation still logjammed in Congress, the worst penalty that Harris could inflict on a poacher was to seize his property and to expel him from the park. In one particularly active week, Harris expelled sixteen men. "The penalties for violation of the rules not being severe," reported Harris, "it has been considered important to make it generally understood that they could not be violated without reasonable certainly that detection with some punishment, or at least inconvenience, would follow."[25]

FOR THE RAWHIDE MEN WHO MADE THEIR LIVING FROM POACHING IN Yellowstone, the threat of "inconvenience" offered little deterrent. Though the West of the late 1880s had tamed considerably from the West of a few decades earlier, there remained plenty of gristle left over from the recent, less civilized past.

The men who poached in Yellowstone had usually made a career out of hardship—hunting buffalo during the heyday, laying track on the transcontinental railroad, fighting Indians in the last years of the Indian wars. Many of the "professional tramps and hard cases" about which Captain Harris complained were fugitives from frontier justice. As Montana's rough railroad towns turned "civilized," the local citizenry became increasingly intolerant of the "obnoxious and disorderly characters" who had been the founding fathers. Vigilante groups often cast such men out, and many of the outcasts drifted naturally toward the notoriously lawless Yellowstone. As Captain Harris complained in one of his reports, "all sorts of worthless and disreputable characters are attracted here by the impunity afforded by the absence of law and courts of justice." For the man without compunction, there were manifold opportunities to make a living, and when weighed against a downside of mere "inconvenience," it didn't take much to entice.[26]

Typical of the breed was a man named Thomas Garfield, whom Harris's men caught trapping beaver in the park. Captain Harris banished him from Yellowstone, but Garfield's days as a menace were not over. The poacher sent a letter to Harris, informing the captain that "[a]s I am ordered out of the Park I am determined to go and take up a homestead on the south boundary line, for the purpose of being a nuisance to the Park and its officers." This was no idle threat. Harris's men later caught Garfield fleeing the scene of a forest fire that he set near the border of the park.[27]

The most notorious of the buffalo poachers was a former sheep-shearer named Edgar Howell, a man equal parts merciless and tough. Like many poachers, Howell used Cooke City, Montana, as his base of operations. Located just across the northeast border of the park, Cooke City offered access to the game-rich valleys of Soda Butte Creek, Slough Creek, and the Lamar River.

For a poacher with the fortitude to ski up Specimen Ridge in winter, the mining town also provided a jumping-off point for the valley of Pelican Creek, where some of the park's harried buffalo sought winter forage among steaming thermal features. Howell was a skilled skier and possessed the survival skills to spend the winter alone in isolated regions of the park—such as Pelican Valley—with only a teepee for shelter. In a poaching career that stretched until 1894, Howell killed an estimated 80

Yellowstone buffalo, single-handedly destroying as much as 40 percent of the bison then alive in the American wild.[28]

IN THE LITANY OF YELLOWSTONE'S INSUFFICIENTLY PUNISHED CRIMES, a stagecoach robbery during the superintendence of Captain Moses Harris was hardly noteworthy at the time. Compared with such criminals as Edgar Howell, who exhibited a ruthless professionalism, the stage robbery was an amateurish, almost comical affair. One of the two robbers was none other than William James—the teamster turned poacher whom Captain Harris had earlier captured and expelled from the park. James, operating out of Gardiner, was attempting to graduate from beaver poacher to road agent. James learned that the army payroll would be passing by wagon through the Gardner River canyon on its way to Mammoth, and with a confederate named Charley Higgenbotham, he set out to steal it.

The two men hid themselves and waited for the evening stage, watching unwittingly as the actual payroll rolled by in a small buggy. When the stage showed up, James blocked the road, yelled "Halt and dismount!" and leveled twin pistols on the teamsters, who complied. The robbers ordered the passengers from the stage, quickly discerning that no payroll was aboard. As consolation, they rifled the passengers' pockets, netting a rather disappointing total of $16. The cash they stole, however, did include two distinctive, foreign coins—an 1811 French piece with the head of Napoleon and an Italian lira with the head of Pius IX. James and Higgenbotham loaded the passengers back up and sent the stage on its way. In the process, Higgenbotham accidentally fired his pistol, startling James so badly that he staggered backward, tripped over a clump of sage, and fell, losing one of his guns.

Two months later, James was camping with two men when he fell victim to traits common in aspiring criminals—the fatal combination of dim-wittedness and verbosity. Alcohol was likely another ingredient in the mix. In a discussion of personal finances, James bragged to his campfire companions that he was "the little son-of-bitch that held up that stage." To prove it, James even displayed the foreign coins.[29]

The campfire companions notified the authorities, and by October, both James and Higgenbotham had been arrested. Despite the clear evidence of their guilt, including the coins, concerns about jurisdiction arose at the trial. The proceedings were held in Montana's U.S. District Court, whereas the crime had been committed in the legal void that was Yellowstone National Park. At a time of harsh frontier justice, the two robbers got away with only a year in prison.

Captain Harris, for one, was outraged—but also resigned. "Although the punishment decreed in these cases appears to be entirely inadequate to the gravity of the offense, yet in view of the uncertainly which seems to exist relative to the administration of justice by the established courts within this reservation, it is perhaps a subject for congratulation that the perpetrators of the robbery were not permitted to escape all punishment." In *Forest and Stream*, Grinnell would note the broader trend toward crime and the broader failure to create meaningful laws for Yellowstone "Instances where visitors have been robbed in the Park have not been very infrequent, and there is now no law to punish the perpetrators of such offenses."[30]

The story of the robbery, however, would not end with the brief incarceration of Higgenbotham and James. The man from whom they had stolen the distinctive coins was a judge from Iowa named John F. Lacey. In a few years' time, Lacey would be a member of the U.S. Congress. He would not forget his personal experience with lawlessness in the Yellowstone National Park.

CHAPTER THIRTEEN

"A Single Rock"

There is one spot left, a single rock about which this tide will break, and past which it will sweep, leaving it undefiled by the unsightly traces of civilization.

—GEORGE BIRD GRINNELL[1]

In 1884, agents for eastern manufacturing companies began to circulate in Montana railroad towns along the Yellowstone River. They were hunting buffalo, two years after the last living animals had been wiped from the Montana plains. Though the buffalo was all but extinct, its gift to the American economy would continue in the form of a grimly final contribution—bones.[2]

The skeletal remains of the once-living herds had two primary industrial applications. Buffalo bones contained phosphate, which when ground and bagged could be sold as fertilizer. Boneblack, another product of bone, was used as a filter in the processing of sugar.

In the early years of the trade in bones, it was not difficult to find them. "Go wherever we might," remembered Smithsonian taxidermist William Hornaday, "on divides, into badlands, creek-bottoms, or on the highest plateaus, we always found the inevitable and omnipresent grim and ghastly skeleton." The old technique of stand hunting aided the bone hunter as much as the hide hunter, conveniently congregating the plunder. Bone hunters simply tossed the bounty into wagons or carts, then hauled it to railheads or steamboat landings. For prairie homesteaders, clearing the ground of buffalo bones was often necessary before it was possible to plow, and farmers jokingly referred to bones as their "first cash crop."[3]

Vic Smith, the famous buffalo hunter, remembered how this new buffalo industry employed "an army of men." For their toil, bone pickers earned $7 per ton. By one estimate, bones generated more money than hides. A single St. Louis facility processed 1.25 million *tons* of bones—the remains of more than 12 million buffalo.[4]

Smith had another memory of the era. Along with the bones of buffalo, the bones of Indians could often be found on the plains, set up on their high burial scaffolds. The scaffolds weakened over time, eventually spilling their contents to the prairie floor. So it was that bones of Indians, so wedded to the buffalo in life, "were gathered up and sold with those of the lordly bison."

A pile of buffalo skulls at the Detroit Carbon Works, circa 1880s.
Courtesy of the Burton Historical Collection, Detroit Public Library.

BY THE EARLY 1890S, THE FATE OF THE LIVING BUFFALO—THE HOUNDED
herds of Yellowstone—had come down to battles on two fronts. In
Washington, D.C., George Bird Grinnell and his allies faced off against
an army of railroad and real estate lobbyists to preserve Yellowstone as a
legal sanctuary from commercial exploitation. In the park itself, the U.S.
Cavalry fought an army of poachers who sought to settle the issue on
the ground. Neither battle was going well.

In February of 1892, Senator Francis E. Warren (of the newly minted
state of Wyoming) introduced a bill to cut away the northern portion of
the park for the purposes of establishing a rail link between Cooke City
and Gardiner. It was the latest in the series of segregation bills that Grinnell and his allies—foremost among them Senator Vest of Missouri—had
been successfully beating back for a decade. The 1892 debate, though,
was about to take an unexpected turn.

Senator Vest was switching his vote. In a remarkable speech on the
floor of the Senate, Vest painted a telling portrait of the politics in robber-
baron America. He began by describing how, for more than a decade, he
had fought against those who sought to "disintegrate the Park and to
mutilate it in the interest of speculators." And indeed Vest, along with
Grinnell, had been fighting the park battle longer than anyone. But
then this:

> *I confess, with considerable humiliation, that I have been*
> *defeated, and I have found, what has been gradually forcing*
> *itself upon my convictions for the last twelve years during*
> *my service in the Senate—that a persistent and unscrupulous*
> *lobby are able to do almost what they please with the public*
> *domain. That portion of the Park cut off upon the north is being*
> *cut off simply because the friends of the Park are unable to*
> *resist the aggressive action of a lobby in the city of Washington*
> *that for years have been endeavoring to force a railroad into the*
> *Park under a charter from Congress in order to sell it for a large*
> *sum to the Northern Pacific Railroad Company.*

Senator George Vest: Grinnell
considered the Illinois senator—
and Boone and Crockett member—
to be his most important ally
in Congress.
*Courtesy of the National Park
Service, Yellowstone National Park.*

Senator Vest's top priority remained the passage of a law creating tough enforcement for park rules. To win such a law, in his view, required the brutal concession of sacrificing a piece of Yellowstone to the railroads. "[T]he fact remains," continued Vest, "that no legislation can be had for the Park until the demands of these people are conceded. It is not a comfortable or a pleasant reflection to a public man to make such an admission, but it is the truth."[5]

Like Senator Vest, Grinnell too was weary after a decade of legislative trench warfare over the park. "I must say that I am sick of this whole Park business," he confided to fellow Boone and Crockett member William Hallett Phillips. In several instances, it appeared that Grinnell might be faltering in his formerly unshakable resolve on the issue of maintaining Yellowstone's territorial integrity. In one editorial he seemed to reflect Vest's view in writing that segregation would "destroy one of the great game ranges of the Park, but this is perhaps better than to have no [enforcement] bill at all." In October 1892, Grinnell would write that cutting away a portion of the park would represent "a grave injury to the reservation," and was "only to be considered as the lesser of two evils."[6]

By December of 1892, though, Grinnell's old vigor had returned in full bore. *Forest and Stream* published a four-page spread titled, "A Standing Menace: Cooke City versus the National Park." It was a blistering indictment of all efforts to shrink the park or to build a railroad through it. Grinnell, having given the issue full consideration, now argued that the idea of segregation as a necessary compromise was "simply jumping out of the frying pan into the fire." In ten numbered sections, Grinnell systematically picked apart every argument made by supporters of a railroad.

In a section headed "Injury to the Park," Grinnell maintained that the "staunchest friends of the bill" were poachers, since segregation would allow them to settle in the middle of the Yellowstone valley. The land to be sacrificed by segregation was a place where "[e]ven buffalo have been recently seen, and with proper protection will soon become abundant."

Grinnell concluded "A Standing Menace" with a direct appeal to Congress: "To all that class who unblushingly place their little interests above a great public interest, who without scruple would inaugurate measures which must lead to the ruin of the National Park, Congress should oppose but one answer, and that should be written in distinct characters on every border of the Park: 'THUS FAR SHALT THOU COME AND NO FURTHER.'"[7]

With "A Standing Menace" as a comprehensive briefing paper and *Forest and Stream* headquarters as what he called the "command center," Grinnell now mounted a furious lobbying campaign to save the park. Grinnell, an outstanding legislative strategist, was also relentless and innovative in the day-to-day tactics of political warfare. In the face of the powerful railroad lobby, Grinnell deployed every asset at his disposal—from deepest grassroots to the pinnacles of Washington power.[8]

Seeking a demonstration of popular support, Grinnell used a *Forest and Stream* editorial to appeal directly to every reader "who would see the Park persevered, for his children and his children's children." He asked subscribers to place a pamphlet version of the "Standing Menace" article "where they will best create public sentiment" and offered to send the reprints "in any desired numbers, post paid, to any address."[9]

The pamphlet included a letter from Theodore Roosevelt, by then a prominent member of the U.S. Civil Service Commission. Echoing Grinnell, Roosevelt encouraged "all public spirited Americans [to] join with *Forest and Stream* in the effort to prevent the greed of a little group

George Bird Grinnell in midlife, at the height of his political influence. *Courtesy of the National Park Service, Yellowstone National Park.*

of speculators, careless of everything save their own selfish interests, from doing the damage they threaten to the whole people of the United States." Roosevelt did not just oppose reducing the size of the park but also believed that "far from having this Park cut down, it should be extended, and legislation adopted ... to punish in the most vigorous way people who trespass upon it."[10]

Along with Grinnell and Roosevelt's individual efforts, the four-year-old Boone and Crockett Club was making its presence known along precisely the insider track that Grinnell had envisioned. A Boone and Crockett meeting and dinner at Washington's elite Metropolitan Club showed the organization at work. At the meeting, Grinnell successfully urged his fellow members to adopt a formal resolution in opposition to any railroad through Yellowstone. For the dinner that followed, the club invited a who's who of Washington players. Roosevelt presided, unveiling the new resolution before sitting down between the Secretary of War and the Speaker of the House. Directly across the table from Roosevelt sat Grinnell, strategically placed beside the Secretary of the Interior.[11]

Roosevelt's flurry of activity as a civil-service commissioner made him a well-known figure on Capitol Hill, and he used his prominence to lobby and testify in opposition to a railroad through the park.[12]

One of the other witnesses to appear before Congress was Captain George Anderson, the third U.S. Army officer to occupy the position of acting superintendent of the park. (After the transfer of Captain Moses Harris from Yellowstone in 1889, the park was administered for a year, unremarkably, by an officer named Frazier Boutelle.) Grinnell hoped that Anderson would preside in the tough-minded, proactive mode set by Harris. Captain Anderson's testimony demonstrated a strong philosophical belief in the mission of protecting the park and its wildlife. He told legislators that segregation would make it impossible to protect the park's game. It would, Anderson argued, eliminate the most critical winter range in Yellowstone, which in turn would cause the destruction of between "25 and 50 percent of the game in the Park."[13]

Despite every effort of Grinnell and his allies, many in Congress remained unswayed. One of the senators advocating segregation was James Berry of Arkansas, who had argued the year before that it was not for the government "to engage in the raising of wild animals," an enterprise that was not a "profitable industry." Berry said that his preferred alternative would be to cut the whole of Yellowstone into 160-acre lots and open it to homesteading, or better yet, to "sell it to the highest bidder and place the money in the Treasury."[14]

In the absence of leadership from Vest, the Senate fell to the potent combination of railroad supporters, real estate speculators, and park skeptics. Grinnell watched in dismay as the Senate—for a decade Yellowstone's last bastion of protection—voted in favor of the bill to cut away a giant swath of the park.

Now the only barrier between the bill and passage was the House, where the railroad lobbyists had always had their way. The Committee on Public Lands voted out the segregation bill with a favorable recommendation. Only a vote by the full House remained—a vote that railroad lobbyists had won on numerous prior occasions. Indeed, past House debates offered a likely preview of what was to come. Representative Lewis E. Payson (the one-time victim of Yellowstone's clumsy legal system) had expressed utter befuddlement at a sentiment that would favor the "retention of a few buffaloes" over a railroad that would lead to the "development of mining interests amounting to millions of dollars, giving profitable employment to perhaps thousands of men."[15]

Payson's viewpoint was echoed by Representative Joseph K. Toole of Montana. "[T]he right and privileges of citizenship, the vast accumulation of property, and the demands of commerce" should not "yield to the mere caprice of a few sportsmen bent only on the protection of a few buffalo in the National Park."[16]

Passage of segregation legislation by the full U.S. Congress appeared unstoppable—until George Bird Grinnell caught a lucky break. To avoid any amendments on the segregation bill (which would have slowed the process of passage), railroad lobbyists sought a House vote under a special floor status called suspension. A lobbyist named P.J. Barr sent a telegram to the governor of Montana, urging him and fellow Montana Democrats to bring "pressure to bear" on Speaker of the House Charles F. Crisp to secure a suspension vote. Someone—Grinnell never revealed his source—slipped a copy of the telegram to *Forest and Stream*.

Truth be told, the effort to secure a suspension vote was a common and perfectly legitimate legislative tactic, but Grinnell seized on it as if the very foundation of democracy had been rocked. He published the telegram on the front page of *Forest and Stream*, along with the headline "Will Speaker Crisp Be Deceived?" The railroad speculators, according to Grinnell, were attempting to perpetrate a fraud on Congress, to pass the bill by "trick and device." For good measure, Grinnell threw in a healthy dose of hyperbole. If the bill passed the House, he argued, the northern portion of the park would be "turned into a howling desert, absolutely without life."[17]

Whatever the merits of these particular arguments, they had the desired effect. Speaker Crisp made the wise decision to stay away from the whole morass. In March, the 52nd Congress expired before segregation supporters could rally another vote. Defenders of Yellowstone and its wildlife had narrowly survived the most significant challenge in the history of the park.

To prevent backsliding, however, was not the same as to make progress. Pushing back the segregation bill in the nation's capitol had done nothing to move forward the urgent cause of meaningful law enforcement in the park itself.

IN THE ABSENCE OF A STRONG LEGAL FOUNDATION, AS GRINNELL knew, nothing was more important than the dedication and skill of the park superintendent.[18] Like Moses Harris, the imposing Captain George Anderson came to Yellowstone with a résumé that included combat experience. Too young to serve in the Civil War (Captain Anderson was the same age as Grinnell—44 in 1893), Anderson had fought against Indians, particularly the Apaches. The ultimate guerilla warriors, the Apaches had schooled the captain in the pursuit of fierce adversaries over hostile terrain. It was experience that Anderson would now put to use in the backcountry battle against Yellowstone's poachers.

Captain George Anderson: Well-suited to his mission, Anderson (standing) combined battle-tested field skills with the ability to operate effectively in Washington's political arena.
Courtesy of the Montana Historical Society.

While both Harris and Anderson brought combat toughness to the position of Yellowstone's top executive, Captain Anderson's background was more diverse—a fact that prepared him for the important political side of his job. Anderson was a graduate of the United States Military Academy (fifth in his class), and his army postings had included a professorship back at West Point, aide-de-camp to the general in charge of the Department of Arizona, and a tour in Europe testing small arms. Hinting at Anderson's political assets, Grinnell would write of him that "few men had so wide a circle of acquaintance."[19]

Unlike the by-the-book Harris, Captain Anderson was loose-laced, quite willing to press the limits of his authority. Consistent with this characteristic, Anderson was a man who enjoyed an occasional nip at the jug. In fact, one of his signature actions in the park was the readmittance of alcohol in the hotels. In support of imbibers, Anderson invoked a number of medical rationales, including the argument that park visitors needed alcohol to help them adjust to the altitude.[20]

Captain Anderson's campaign against poachers had the benefit of building on the solid foundation built by his two army predecessors. For example, Anderson continued the use of winter ski patrols, an activity now supported by a network of five "snowshoe cabins" at strategic points in the backcountry. These cabins made it possible for patrols to stay in the field for extended periods.[21]

Anderson realized that the vastness of the Yellowstone territory made it essential to deploy more men. He requested and was granted an additional troop of cavalry, though the army declined the captain's request for $900 to hire a second scout. When patrolling the park, Anderson ordered his men to seek out the least-traveled paths, sniffing out the hidden pathways of the poachers.[22]

Captain Anderson credited Moses Harris for having "put in motion" most of the law-enforcement techniques at use in Yellowstone. Certainly Harris deserved enormous credit, but Anderson put his own stamp on park protection. His most important contributions were an agility and creativity of action—combined with keen political savvy—that would prove critical for the era in which he served.

Anderson understood that public opinion, if properly mobilized, could be a potent ally in his work. Thus the captain greeted the widespread practice of writing graffiti on park formations with a new form of punishment, rather elegant in its simplicity. "I ordered that every man found writing his name on the formations should be sent back and made to erase it." As they scrubbed away at their handiwork, reported Anderson, transgressors could listen to the "taunts and gibes" of their fellow tourists.[23] Such punishment was also a useful public display that the army meant business.

As in any conflict, Captain Anderson knew that his war on poachers required timely and accurate intelligence, and he worked for years to develop a sophisticated network of allies and informants. Civilian law-enforcement officials from surrounding communities were one ongoing source of information. One of Anderson's frequent correspondents was William McDermott, the U.S. marshal in the wide-open mining town of Butte, which also happened to be home to a number of taxidermists. In one letter, Marshal McDermott warned Anderson that a poacher had "corralled" two buffalo somewhere "in or near the Park," and that he "intends to kill them as soon as sales for heads can be arranged." Wyoming's gamekeeper, John Krachy, wrote Anderson with an offer to work together in the "protection of the game," noting that Wyoming's lawbreakers "hunt in the Park."[24]

Anderson also relied on a group of guides who operated in and around the park. One such man, E. W. Robins, wrote the captain in support of a proposal requiring guides to register with park officials. Robins noted the helpful role that guides could play in enforcing "strict observance" of park rules, "especially when in the more unfrequented portions of the Park where there are no soldiers to watch." Outfitters were another good source of information. G. W. Marshall, proprietor of a Bozeman store with the descriptive name of Montana Armory, sent Anderson a letter with details he had overheard of a plan to capture game in the park, "especially buffalo."[25]

Captain Anderson corresponded frequently with George Bird Grinnell, who made the politically wise move of admitting Anderson to Boone and Crockett. The two men became both close personal friends and coconspirators in the battle to save the buffalo and the park. Grinnell's letters to Anderson contained not only the latest gossip from Washington, D.C., but also information from *Forest and Stream*'s field reporters, who were frequently dispatched to the park. In an episode that underscores the journal's influence, Secretary of the Interior John Noble wrote a letter to Anderson with a recent *Forest and Stream* article as an attachment. The article discussed small bands of buffalo living just outside of the Yellowstone boundary. Secretary Noble asked Anderson to "ascertain if these bands may not be brought within the Park and retained there until they feel greater security."[26]

One of Anderson's best sources of intelligence fell under the general category of "concerned citizens." *Forest and Stream*'s two decades of constituency building had made its mark on the country, and when tourists and others saw evidence of poaching, many were angry enough to act. F. Lusk, whose letterhead identified him as a "shorthand reporter," wrote Anderson to inform him that a buffalo killed in the park was being shipped to a market in Chicago. An anonymous correspondent, "Veritas," sent Anderson a ten-page letter with a wealth of details concerning both buffalo hunting in the park and a network for selling heads that included operatives in Pittsburgh and New York City. "I have been a reader of *Forest and Stream* for many years," wrote Veritas, "and have always taken a live interest in the endeavors of this sporting paper on behalf of the protection of game."[27]

With the information from his network of sources, Anderson and scout Edward Wilson developed a sort of most-wanted list. The three most notorious poachers were E.E. Van Dyck, Charles Patterson, and Edgar Howell. All of these men used Cooke City as the staging ground for hunting and trapping forays into the park.[28]

In February of 1891, one of Captain Anderson's informants passed along information that a Livingston taxidermist had just mounted three or four heads—reportedly provided by E.E. Van Dyke. Anderson responded immediately, dispatching three patrols to the Lamar Valley and Soda Butte regions where Van Dyke was known to operate. A few days later, the patrol led by Edward Wilson returned to Mammoth with a report. Van Dyke had been located near Soda Butte, but the cavalry contingent had been unable to approach him. The poacher, according to Wilson, spent the daylight hours with a field glass "on the highest points in the neighborhood," a vantage that allowed him to scan every potential approach up the broad valley. Any attempt to capture him would have been futile. Wilson told Anderson that the "only way to get him" was for one man to go alone. The approach would have to be made from behind, requiring a perilous winter crossing of Specimen Ridge. It would also have to be made at night. Wilson, of course, volunteered.

According to Captain Anderson, the scout set out a 9 o'clock in the evening in "as bad a storm as I ever saw." Wilson skied forty miles, including a nighttime ascent of Specimen Ridge. As the sun rose, Wilson set up a concealed position that afforded a view of Van Dyck's camp. Then he settled in to wait. Not until twilight did the cautious Van Dyck finally made a move for the open valley below him. Wilson watched as the poacher skied down to check beaver traps in the Lamar River. It was the hard evidence that Wilson needed. Now the scout settled back to catch a quick few hours of sleep in a frigid, fireless camp.

In the darkness before dawn, Wilson moved on Van Dyke, gliding silently on skies. First light was just breaking as Wilson crept into the poacher's camp. Not imagining that anyone could invade his mountain-top lair, Van Dyck slept soundly away. In fact so deep was Van Dyck's slumber that Wilson decided to tarry a little longer. Using a Kodak camera that looked like a small leather box, Wilson actually photographed the sleeping poacher! Then he woke him up and put him under arrest. The photograph of a man sleeping seems unlikely to have had much evidential value. More likely Wilson intended it for the amusement, likely to have been considerable, of his comrades back at Mammoth.[29]

Wilson marched Van Dyck across the forty miles back to Mammoth, there to encounter the perennial quandary of law-enforcement officials in the park. Having gone to extraordinary lengths to capture lawbreakers, how then to punish them? The tools at Captain Anderson's disposal were the same lame penalties applied by his predecessors—confiscation and eviction.

As for confiscation, Anderson seized so much contraband that he ran out of storage, resorting to a massive bonfire to make room. The punitive value of confiscation, never great, was diminishing over time. Poachers, aware that seizure of their property was one of the few measures that the army could take, made it a point to carry inexpensive equipment and supplies. If they were captured, the confiscation of their plundered heads and furs represented the only real loss. The sacrifice of some cheap equipment, by contrast, was an acceptable cost of doing business. On the rare occasions when more valuable properties were confiscated, Anderson sometimes put some of them to use. One self-admitted elk poacher sent a letter to the Secretary of the Interior to complain that Captain Anderson's son was riding his confiscated horse. More typically, though, if poachers happened to be caught with something of value, they quickly discovered a loophole that allowed them to win the return of their possessions. As the captain complained bitterly, "every bit of property found on such men is at once claimed by their partners and confederates, on real or fraudulent bills of sale." The army usually relented in the face of such claims. The end result, in Anderson's words: "[C]onfiscation if made hurts the transgressors very little."[30]

Frustrated and eventually enraged at his inability to make punishment hurt, Anderson began to get creative. Under Captain Anderson, for example, the penalty of expulsion from the park assumed a new flavor. Anderson knew that merely removing poachers from Yellowstone was little deterrent, so rather than marching poachers to the nearest park boundary, he began a policy of marching them to the most distant boundary. Instead of enduring "inconvenience," poachers and other lawbreakers now found themselves on extended, forced marches.[31]

Anderson no doubt derived some small satisfaction from marching poachers for scores of miles over difficult terrain. But still, he knew that most would come back. E. E. Van Dyke was a case in point. There was a reason that scout Edward Wilson had placed this particular poacher on his most-wanted list. Van Dyke had been captured and expelled from the park a year earlier, only to return. "[S]imple removal," reported Anderson, "had absolutely no effect on such characters."[32]

What was necessary, as George Bird Grinnell had argued for more than a decade, was real penalties—and that meant the ability to lock poachers behind bars. It was in this domain that Captain Anderson undertook the most controversial actions of his Yellowstone superintendence. Despite his complete absence of legal authority, Anderson threw Van Dyke in the guardhouse. As a thin gauze of justification, Anderson claimed that he needed to hold the poacher while "awaiting the instructions of the Secretary of Interior." "[F]or some reason," Anderson later related gleefully, the secretary's instructions "were very slow in coming."

This probably had something to do with the fact that Anderson waited a month before notifying the Department of the Interior that he had taken a prisoner.[33]

When instructions did come (the day after Anderson told Washington what he was doing), they reminded the captain that the Congress had given no authority to hold prisoners—even those guilty of destroying the last few buffalo alive in the wild. Anderson, for his part, was unapologetic. "[T]his imprisonment of a month," he told the secretary, had been more effective than "all the other arrests made since the Park was established."[34]

There can be little question that Captain Anderson and the U.S. Army were doing everything within their power—indeed an expansive interpretation of their power—to stop poaching in the park. Without the support of a true legal framework, however, poaching would continue. Even as the number of buffalo fell precipitously, the human population in the communities around the park grew rapidly, increasing the number of men who took a shot at poaching. As for poachers, the fundamental calculus remained compelling: The reward was great and the risk was small. The ever-increasing scarcity of the buffalo served only to drive up the price. In the spring of 1894, Anderson reported to the Secretary of the Interior that mounted buffalo heads were selling for $500 or more if the customers were "rich and anxious." In London, according to the captain, a head had sold for £200 ($1,000). A Wyoming-based poacher named John Ching was brazen enough to advertise his services. In *American Field*, Ching invited "Gentlemen who want *Bison*" to contact him.[35]

In 1894, a Livingston taxidermist named Earl B. Wittich traveled to Minneapolis by train, engaging in a long conversation with the man sitting beside him, a St. Paul attorney named E.S. Thompson. When Thompson learned that Wittich was a taxidermist, he began to ask about the availability and price of various mounts, eventually inquiring if it were still possible to obtain a "prime" buffalo head. "Certainly," replied Wittich, "and as many more as you wish." When Thompson expressed surprise, Wittich told him that he had "actual knowledge" of 181 buffalo killed within the park during the past winter, many of which he had mounted himself.

Attorney Thompson collected taxidermist Wittich's contact information and went home, but then his conscience got the better of him.

He wrote letters to both the Secretary of the Interior and Captain Anderson providing a full report of the encounter, even suggesting a covert strategy for capturing Wittich's suppliers. (It's not clear in what manner Anderson followed up.) Wittich, according to Thompson, claimed to regret the "very probable extinction" of the buffalo. The taxidermist's bottom line, though, was the same as the hide hunters who preceded him. As Thompson described it, "in view of the fact that the killing was certain to be made, someone was sure to make big money and he thought that it might just as well be himself as anyone else."

Despite his apparently recent conversion to the cause of conservation, lawyer Thompson had a clear sense of where things were going. "There are but a very few animals now surviving," he wrote to Anderson, and "each one can be killed but once."[36]

KILLED ONCE, BUT HARVESTED TWICE: ONCE BY THE HIDE HUNTERS; once by the bone pickers. By 1894, though, the bones too were gone, and there were no more herds to replace them. The American buffalo, having sustained the inhabitants of the Great Plains for millennia, had finally given all that it could.[37]

CHAPTER FOURTEEN

"For All It Is Worth"

I am going to use the recent unfortunate slaughter of buffaloes for all it is worth for trying to get legislation through.

—Civil Service Commissioner
Theodore Roosevelt in a letter
to Captain George Anderson[1]

In the fall of 1893, poacher Edgar Howell pulled out of Cooke City for his winter hunting grounds in Yellowstone. He took a partner, a man with the incongruous name of Noble, and a dog, a mixed-breed shepherd with all but the nub of its tail chewed off. Like his dog, Howell was rough around the edges. A reporter who saw him that winter described the poacher as "picturesquely ragged." He was dirty and unshaven, his hair flowing back in a greasy "drake's tail," his beard occasionally hacked off with scissors when it got in his way. Howell was no more than average in height, and the words used to describe his movements—*slouchy, stooped, loose-jointed*—fail to convey his underlying grit.

Even the men who pursued Howell, men who despised him for his illicit trade, could not help but feel a grudging respect. "Howell," said the reporter, "was in his brutal and misguided way a hero in self-reliance." The West admired a man who could go up against the elements and endure, and none exceeded Howell in his ability for sheer survival. To elude the cavalrymen whom he knew patrolled the Soda Butte and

Lamar Valleys, Howell skied over the fearful heights of Specimen Ridge. Portions of the ridgeline trail soared above 8,000 feet, and winter snows were commonly ten feet deep. Not only did Howell pull his own weight up the steep slope but he also towed a ten-foot toboggan larded with 180 pounds of supplies—the same type of sled that the army had abandoned as infeasible on far flatter terrain.[2]

Howell established his winter camp in a remote area north of Lake Yellowstone known as Pelican Valley. As the poacher knew, it was the ideal winter refuge for buffalo—a broad valley divided by a shallow creek and sheltered on both sides by timber and rising hills. The valley also contained a number of thermal features, "hot ground" that helped to keep the snows at bay. Together with his partner Noble, Howell set up a teepee at the confluence of Pelican and Astringent creeks. In the process of establishing their base of operations, Howell and Noble had a falling out. The precise cause of their altercation is unclear, but it ended with Howell booting Noble out of camp. Noble headed south toward Jackson Hole, leaving Howell to ply his trade in frigid solitude.[3]

Not that he minded, for Howell was a man fully accustomed to taking care of himself. In the middle of the Yellowstone wilderness, Howell suffered an accident that might have proved fatal to a lesser woodsman: He broke a ski, robbing him of the only means of transport that could carry him over snow that was deeper than a man's head. To repair this vital piece of equipment, Howell used his ax to hew a thin, five-foot slab from a fir tree. This he fashioned into a new tip for his broken ski, nailing it to the old stub and then smoothing out the repair with resin from the fir.[4]

Howell's plan for the winter of 1893–1894 was straightforward and practiced. Using his skis to cover wide swaths of Pelican Valley, the poacher would scout the rolling creek bottom for signs of his prey. Once the buffalo were located, winter made easy the job of killing. The heavy buffalo could barely move in the deep snow, allowing Howell to ski up beside the panicked, floundering animals—then shoot them point-blank. Once he dispatched the animals, Howell cut off the trophy heads, wrapped them in gunnysacks, and hung them in high trees to protect his plunder from scavenging animals. The heads were too heavy to remove from the park in wintertime. Instead, Howell planned to return in the

On the run: This haunting photo, taken in Yellowstone in the mid-1880s,
shows one of the few small herds of buffalo left alive in the wild.
Courtesy of Archives and Special Collections, University of Montana-Missoula.

spring with a pack horse. He had done it all before, killing Yellowstone
buffalo by the score. Now, with everything again in place, Howell started
hunting.

As Captain George Anderson hunted Edgar Howell in the
winter of 1893–1894, he did so without his most trusted and potent
weapon. In the summer of 1891, scout Edward Wilson left Mammoth
on a routine mission, never to return. It would be a year before his body
was discovered, barely a mile away, the apparent victim of a suicide. In
an oblique reference to the event, Captain Anderson wrote that Wilson
was "brave as Caesar, but feared the mysterious and the unseen."[5]

Devastated, Anderson found himself without the services of the very
man who had proven the viability of winter patrols in the park. In his
1891 report, the captain wrote that "it will be quite impossible to replace
him." Fortunately for Anderson and for Yellowstone, the new scout he
hired that year possessed all the moxie of his predecessor. His name was
Felix Burgess, and like Anderson, he was a veteran of the Indian wars in

the American Southwest. Though an expert skier, Burgess often walked with a limp. In his military days, he had been captured and tortured by Indians, who, according to Burgess, made him "put [my] foot on a log, then amused themselves by cutting of [my] toes, taking two off clean and nearly cutting off the great toe." The "great toe" now had poor circulation, turning painful and black whenever Burgess engaged in winter travel through the frigid Yellowstone backcountry.[6]

Burgess was on just such a mission in January of 1894 when he discovered the first trace of Howell. Captain Anderson suspected that Pelican Valley would be a target of poachers after leading a fall patrol that found "old signs of bison." From his informants in Cooke City, the captain also knew that Howell was on the move, though his precise whereabouts in the vast park were unknown. As Burgess scouted the east park in January, he came across a set of tracks made by skis and a heavy sled. According to Anderson's report, the tracks were "old and could not be followed." Still, Burgess would have stayed in the area and scouted further had it not been for a piece of bad luck. He broke his ax, and the loss of such a vital tool could not be ignored. Most critically, of course, an ax was often necessary to obtain firewood in the snow-blanketed terrain (not to mention such unanticipated emergencies as ski repair). Disgusted, but without alternatives, Burgess headed home.[7]

Back at Mammoth, Captain Anderson listened intently to his scout's report. Only a poacher would travel the park in winter—and only a man such as Howell had the fortitude to pull a sled. Howell, he now strongly suspected, was operating somewhere in the remote, snowbound stretches of Pelican Valley.

GEORGE BIRD GRINNELL KNEW THAT HE HAD BEEN SAVED BY THE bell. In fighting against segregation legislation in the spring of 1893, the lucky break of obtaining a railroad lobbyist's "secret" telegraph to the Speaker of the House had provided the opening he so desperately needed. When Grinnell published the telegraph on the front page of *Forest and Stream*, he successfully muddied the waters—then ran out the clock on the 52nd session of the U.S. Congress.

Now, though, in the fall, it was the railroad lobbyists who had the opening. A nationwide financial crisis had broken out, the Panic of

1893, and to address the crisis, President Grover Cleveland called Congress into special session. The railroads, ever ready, pounced on the unexpected session as an opportunity to press two bills aimed at Yellowstone National Park. One bill was virtually identical to earlier efforts to cut away the portions of the park to the north of the Yellowstone River. The other attempted to win congressional approval for a railroad—this time an electric railroad—from Cooke City to Mammoth.[8]

Realizing how narrowly he had averted disaster only months before, Grinnell once more climbed the ramparts and set about the task of mounting a defense. As he had done in the earlier session of Congress, Grinnell wrote to Captain Anderson and told him that "it would be a very good thing if you could induce the Secretary to order you here so that you can be on hand to fight the Park Battles." The commanding, well-connected Anderson had been an effective witness at congressional hearings, and Grinnell wanted all hands on deck. "I do not feel very cheerful about the prospects," reported Grinnell to Anderson, "but I should like to think, when the battle is over, no matter how it may go—whether for or against us—that I had faithfully done my part in fighting for the reservation!"[9]

As it turned out, the cumulative effects of a decade of political battles had finally paid off—at least on one front. Weary of the Yellowstone railroad issue, House and Senate committees voted down both the segregation bill and the electric railroad bill. There would be a few scattered efforts at park segregation in the years to follow, but none would gain traction.[10] The battle against the railroad had been won.

Grinnell was elated, though also stunned. "[W]e struggle and sweat and fight for fifteen or twenty years or more, without apparently making any progress," he later described it, "and then of a sudden the things we were working for come to pass."[11]

Never resting, Grinnell and Senator George Vest prepared immediately to shift from defense to offense. Believing that their victory over the railroad signaled a new depth of congressional interest in Yellowstone, Senator Vest reintroduced his long-stymied enforcement legislation for the park. Surely the moment was at hand.

But to the utter dismay of park allies, Congress yawned. Though the park had finally been liberated from the railroad, the Vest enforcement bill failed to find support. The limits of congressional interest, it appeared, had been surpassed. Grinnell's rage over what he called Congress's "criminal negligence" poured out on the pages of *Forest and Stream*. Park enforcement legislation, he argued bitterly, "has in it nothing to excite the enthusiasm of the politician . . . nothing but the public good." Congressional inaction was a "disgrace to every citizen."[12]

What Grinnell had come to realize by the winter of 1894 is that there is no crisis more pernicious than the slow-motion disaster. Human nature, and with it the American political system, are geared to respond to the immediate, the proximate, and the tangible. The gradual, the distant, and the abstract are the enemy of action. When it came to the destruction of the buffalo, Grinnell had succeeded in making people aware that it was happening—he had even succeeded in making Americans care that it was happening.[13] But there remained a gaping chasm between concern and action.

What Grinnell needed was a stick in the nation's eye—an event so palpable, so painful as to incite collective outrage. In Yellowstone National Park, though Grinnell could not yet know it, the event he needed was about to present itself in the personage of Edgar Howell.

By February 1894, Edgar Howell already could count the winter as a lucrative season. In the secluded haunts of Pelican Valley, the poacher had killed at least six and possibly fourteen or more buffalo. The majority of them were cows and calves, less valued by the trophy market, but Howell knew that the broader condition of scarcity would reward him well for his toil. Plus, the poacher knew there were still a few more buffalo for the taking. There had been no sign of cavalry all winter, and given the choice, Howell would have remained in the remote valley without interruption. Events, though, had forced his hand—he was running out of supplies. Reluctantly, Howell made the decision to undertake the arduous winter journey into Cooke City.[14]

There was a reason that Captain Anderson had not yet ordered his men back into Pelican Valley. With Felix Burgess's discovery of a single track as his only clue, Anderson lacked the information he needed to order men into dangerous country. The dangers of winter travel were significant, and with only the knowledge he possessed as of February

1894, the captain knew that he was aiming at the void. Desperate for more information, Anderson mobilized his broad intelligence network.

Anderson knew that one of his most important assets was the army's winter "snowshoe cabin" near the distinctive formation of Soda Butte. Soda Butte Station sat in prime territory astride the most traveled valley pathway between Cooke City and the winter hunting grounds in the park. Anderson doubted that an experienced poacher like Howell would be foolish enough to march up the Soda Butte Valley, but he would likely pass nearby. At about the time Burgess was stealing back across Specimen Ridge on his way to Cooke City, the captain ordered his men at Soda Butte to broaden their patrols to the surrounding mountains. Within a few days, in the mountains near the Soda Butte Station, Anderson's men found the recent tracks of a man on skis, towing a toboggan. The tracks trailed down from the mountain, leading clearly in the direction of Cooke City.

The soldiers notified the captain, who quickly pressed his informants in Cooke City. Edgar Howell, Anderson learned, had indeed come in. The poacher had not stayed long, reprovisioning himself and then heading back into the park.

Now Captain Anderson had the information he needed to make a move. "Calculating the time when [Howell] should be in the bison country, I gave Burgess an order to repeat his trip." Ribbing his scout over the earlier equipment mishap, the captain ordered him to come back "with a whole ax and a whole prisoner, if possible." Still, Anderson warned Burgess against taking "risks or serious chances." The captain, not a man prone to exaggerate, believed Howell was a "desperate character" who would "resist arrest even to the point of taking life."[15]

One measure of Captain Anderson's dedication to ridding the park of poachers (and Congress's corresponding disinterest) can be found in the funding for Felix Burgess's March mission into Pelican Valley. There was no government money available, so Anderson funded the expedition himself—$130.58 from his own pocket.[16]

On March 11, 1894, Felix Burgess and a private named Troike[17] departed the Lake Hotel (on the northern end of Lake Yellowstone) in what Anderson described as "a severe storm."[18] The Pelican Valley region in winter was like a rolling sea of snow, and even with skis, the labor of breaking trail would have been intense. Burgess and Troike likely traded off, one man clearing the trail for a time while the other enjoyed the relative respite of gliding behind in his tracks. It took the men a full day to ski back to the site where Burgess had seen the tracks in January—probably along a small Pelican tributary called Astringent Creek. There they set up their camp, settling in as best they could through the frigid night.

According to an account given by Burgess, the next morning the two men "got out early and hit the trail not long after daybreak."[19] They were barely out of camp when they stopped suddenly in their tracks, staring up at a macabre sight. Dangling from the high branches of a pine tree were the heads of six buffalo. Beneath the tree were ski tracks, though they were too old to follow. Burgess didn't need tracks now, though. He had the scent. Howell would have placed his cache in a place where he could easily find it, and the most logical path followed directly down Astringent Creek into Pelican Valley. Below them in the valley, they now knew, was their quarry.

The downhill run along Astringent Creek allowed the two men to ski quickly, tempered only by the need for caution in approaching the notorious Howell. The sense of foreboding grew deeper when, around noon, they crossed fresh ski tracks. The coulee cut by Astringent Creek deepened as it approached Pelican, shrinking their field of vision. A howling wind also blew, masking all but the most penetrating sounds. Suddenly they rounded a bend and found themselves staring at the teepee that was Howell's camp.

Fresh tracks surrounded the teepee, including a set of ski tracks that led off toward the south. As the two men moved cautiously through the camp, Burgess described what happened next. "I heard the shooting, six shots." With a mixture of trepidation and exhilaration, Burgess and Troike realized that Howell was just over the ridgeline. Quickly they skied up the hill, hugging the tree line for cover.

As they peered over the ridge into the broad valley below, they spotted Howell at a distance of about 400 yards. Around him lay the forms of five dead buffalo, "slaughtered in the deep snow in which they had wallowed and plunged." Howell faced his pursuers across open ground, the deep snows burying every bit of potential cover. Preoccupied with his kill, however, Howell failed to notice them. "His hat was sort of flapped down over his eyes," remembered Burgess. "He was leaning over, skinning on the head of one of the buffalo."

Burgess and Troike must have briefly discussed their next step, though the plan that emerged was hardly complex. "I thought I could maybe get across without Howell seeing or hearing me," said Burgess, "for the wind was blowing very hard."

Two dangers faced Burgess in his decision to charge across the open plain—one he could see, but the other he could not. The obvious threat was Howell's rifle, as Burgess recalled, "leaning against a dead buffalo, about fifteen feet away from him." With a rifle, Howell could shoot a man down at a range of 200 yards or more. Burgess's only weapon was a .38-caliber army revolver—a gun whose effective range was probably no more than 20 yards. It was a horrible mismatch, and if Howell happened to look up as Burgess crossed the open ground, the scout would be dead. (Remarkably, Private Troike was unarmed, which explains why he stayed back in the cover of the timber.)

Given how critical it was for Burgess to approach unnoticed, the second threat he faced was equally deadly. Curled up behind one of the buffalo carcasses was Howell's dog, the bobtailed shepherd. One bark would alert the poacher to the charging Burgess.

Wholly ignorant of the dog, Burgess gave a mighty push on his single ski pole, propelling himself forward into open ground. "I started over from cover, going as fast as I could travel." For 200 yards Burgess made rapid—and silent—progress. But then an unexpected obstacle appeared. "Right square across the way I found a ditch about ten feet wide, and you know how hard it is to make a jump with snowshoes on level ground." It was too late to do anything but plow forward. "Some way I got over." Even as Burgess grunted up the steep bank of the ditch, Howell still failed to look up. Nor did the dog. "The wind was so the dog didn't notice it," remembered Burgess, "or that would have settled it."

"I ran up to within fifteen feet of Howell, between him and his gun, before I called to him to throw up his hands, and that was the first he knew of anyone but him being anywhere in that country." Burgess recalled that Howell, at first, "stood stupid like," then the scout told the poacher to drop his skinning knife.

Cavalrymen and civilian scout Felix Burgess ski toward Mammoth
with captured poacher Edgar Howell.
Courtesy of the National Park Service, Yellowstone National Park.

Howell dropped the knife. "All right," he said, "but you would never
have got me if I had seen you sooner."[20]

Burgess agreed. "I expect probably I was pretty lucky," he later admitted. "Everything seemed to work in my favor."[21]

Burgess called to Troike, who quickly skied down.

Howell now turned his anger on his dog, outraged that the animal
had failed to give warning. "Howell was going to kill the dog," according to Burgess, but "I wouldn't let him."

As they prepared to march (ski) Howell back to the guardhouse at
Mammoth, Burgess and Troike observed firsthand that Howell's reputation for toughness was well deserved. In the midst of a Yellowstone
backcountry winter, the poacher had no shoes, "only a thin and worthless pair of socks." Around his socks he wrapped gunnysacks. To attach
his jury-rigged skis, Howell stuffed his feet into "a pair of meal sacks he
had nailed onto the middle."[22] Then he bound up the whole contrap-
tion with thongs.

The Yellowstone National Park Archives identify this photo as
"Soldiers with Howell," who is probably the man on the right. The two
in the center are likely Burgess and Troike. The man on the left is
believed to be *Forest and Stream* reporter Emerson Hough. Curled up
in the snow is Howell's dog.
Courtesy of the National Park Service, Yellowstone National Park.

By late morning on March 12, Burgess and Troike—with Howell out
in front to break trail—were skiing toward the hotel at Lake Yellowstone.
Burgess intended to pass the first night at the hotel, then make his way to
Mammoth via the winter station at Norris.

AT 9:30 PM ON THE EVENING OF MARCH 12, THE TELEPHONE RANG
at the U.S. Army's Mammoth headquarters. Winter telephone service
between Lake Yellowstone and Mammoth was tenuous at best, the connecting wire requiring constant attention by a lineman named Snowshoe Pete. That night, though, scout Felix Burgess's call came through,
and the news it carried left the captain "positively jubilant," according to
a reporter who was there. "He couldn't sit still, he was so glad."[23]

Certainly the capture of the notorious Howell was cause for celebration. Despite the high drama of the confrontation between Howell and
Burgess, however, there was in some ways little in the poacher's arrest to
distinguish it from other successes of Anderson and his men. Most glar-

ingly, of course, there was still no law under which Anderson could effectively punish the poacher his men had captured. The captain planned to confiscate Howell's property and to "sock him just as far and as deep into the guard-house as I know how." But within a few weeks, as Anderson said at the time, "the Secretary of the Interior will order me to set the prisoner free."[24] And so the vicious cycle would repeat, each time with fewer buffalo than before.

Yet this time it would be different. When Anderson hung up the phone after talking to his scout, there was a guest at Mammoth with whom he could share the news. The guest was Emerson Hough, a reporter from *Forest and Stream* magazine, dispatched to Yellowstone by George Bird Grinnell as part of a broader effort to determine the status of the buffalo in the park. As Grinnell explained it, the imperative for firsthand information arose from the combination of "conscienceless money-hunters in the United States Congress" and "conscienceless head-hunters" in the park. "Now, if ever," wrote Grinnell in a *Forest and Stream* editorial, "the time has come to sound a call which shall awaken anew public interest."[25]

Beyond Grinnell's greatest expectations, his expedition had stumbled on a pot of political gold. Within twelve hours of Howell's capture, Grinnell was reading about it in a telegram sent from Mammoth by Hough. Grinnell seized instantly on the incident, boarding a train for Washington the next morning to rally congressional allies.

As for Anderson, he immediately dispatched a ski patrol from Mammoth to collect evidence from the Pelican Creek site of Howell's massacre. (With the responsibility for guarding his prisoner, Burgess and Troike had been unable to transport the heads and hides.) Reporter Emerson Hough would accompany the outbound soldiers, as would another vital player, respected wilderness photographer F. Jay Haynes.[26]

Two days later, at Norris Station, the outbound patrol from Mammoth met up with the inbound Burgess, Troike, and their prisoner. Both groups spent the night at Norris, giving Emerson Hough the opportunity to gather interviews that he would later fold into a detailed account for *Forest and Stream.*

The reporter was both fascinated and outraged by his encounter with Edgar Howell. The poacher's disposition was "chipper and gay," according to Hough, and Howell "seemed to enjoy the square meals of captivity." In one breakfast, the poacher consumed twenty-four pancakes. While a full stomach may have contributed to Howell's buoyant mood, more significant was the fact that he took great pleasure in torturing his captors with the knowledge that it was the captive who still held the upper hand. The poacher, for example, taunted Hough and the soldiers with the compelling math of his illicit trade. He bragged that he would have earned more than $2,000 from his buffalo heads—more than ten times the annual salary of an army private. As for his losses (he knew that his property would be confiscated), Howell estimated that the value of his equipment was $26.75. "Yes," Howell continued, "I'm going to take a little walk up to the Post, but I don't think I'll be there long." As for his future plans, Howell said coyly that "I may go back into the Park again later on, and I may not."[27]

While Howell, on a full stomach, toyed with his keepers, the man who had risked his life to capture him was suffering greatly. By the time he reached Norris Station, Felix Burgess's old foot injury "had frozen again on this trip, and was in bad shape." Emerson Hough watched him ski into the station, "limping very badly." According to Hough, the scout said nothing about his injury but went immediately to the fire, "showing a foot on which the great toe was inflamed and swollen to four times its natural size. The whole limb above was inflamed and swollen and sore, with red streaks of inflammation extending up to the thigh," continued Hough. "How the man ever walked I cannot see."

The next day, Burgess would forge on with his prisoner, delivering Howell to Mammoth after a final push across more than twenty frozen miles. Good to his promise, Captain Anderson threw the poacher in the guardhouse, there to await the inevitable release order from Washington.

Almost immediately after Burgess arrived, the post surgeon amputated his toe—narrowly saving the scout from death by gangrene.

GRINNELL BROKE THE STORY OF THE HOWELL INCIDENT IN THE March 24 edition of *Forest and Stream* with a brief news bulletin wired from Emerson Hough: "Scout Burgess has captured Howell, the notorious Cooke City poacher with ten fresh buffalo skins, on Astringent Creek, near Pelican." Along with this bulletin, Grinnell ran a front-page

Howell's plunder: Soldiers pose with the grisly remnants of
Edgar Howell's work.
Courtesy of the National Park Service, Yellowstone National Park.

editorial titled "A Premium on Crime." Yet again, the *Forest and Stream*
editor excoriated Congress for its ongoing failure to pass enforcement
legislation. At the same time, Grinnell emphasized the significance of
Howell's capture, calling it "unquestionably the most important that has
ever been made in the National Park."[28]

Grinnell, having hurried to Washington at the first news of Howell's
arrest, recognized immediately the political potential of the incident. In
the Senate, Grinnell's old ally George Vest would continue his role as
what Grinnell called "the watchful guardian in Congress."[29] In the
House of Representatives, meanwhile, Grinnell found a new ally in the
form of Iowa Congressman John F. Lacey.

Congressman Lacey, it will be remembered, had had his own intimate experience with lawlessness in Yellowstone National Park when
road agents stole his two coins in the botched 1888 stage robbery. Then,
Lacey watched in frustration as the thieves received light sentences—

primarily because of the uncertainty over the legal status of the park. Now, six years later, Lacey was a member of Congress (and another well-placed member of Boone and Crockett). Only two weeks after Howell's capture, Lacey—with Grinnell's active encouragement—introduced enforcement legislation that came to be known as the Lacey Bill. Senator Vest pressed parallel legislation in the Senate. Grinnell had his vehicles.

Now Grinnell reached for every weapon at his disposal, drawing on an armory amassed in two decades of fighting for wildlife and the park. *Forest and Stream*, as always, provided the central platform. Grinnell's themes were familiar, but the Howell incident provided a gruesome exclamation point, a tangible case in point to arguments that had failed, in the past, to result in action.

Emerson Hough continued to provide commentary from the park, filing detailed dispatches after returning from the expedition to investigate the Pelican Creek massacre site. *Forest and Stream* subscribers read accounts of how Howell had taunted his captors with the knowledge that any punishment would be slight. Perhaps most impactful were stark photographs snapped by Jay Haynes. For the first time, Americans could actually *see* the gory handiwork of poaching—masses of dead buffalo lying slain against the deep winter snow.[30]

Grinnell circulated reprints of the articles through the halls of Congress, and lest any member failed to connect the dots, he urged every *Forest and Stream* reader to "write his Senator and Representative in Congress, asking them to take an active interest in the protection of the Park." *Forest and Stream* even circulated petitions around the country. The result, according to Grinnell, was that "thousands upon thousands of these, fully signed, came into Congress from all quarters."[31]

A week after Congressman Lacey introduced his park enforcement bill, the full House voted to pass it.[32] In the Senate, meanwhile, Grinnell marshaled the full force of his heavy-hitting Boone and Crockett brethren. Theodore Roosevelt, in full righteous outrage, wrote to congratulate Captain Anderson on his work and to vent his spleen about the "infernal scoundrel" Howell. "A man like that ought to be sent up for half a dozen years." In a meeting with William Hallet Phillips, Roosevelt expressed the opinion, perhaps hyperbolic, that Anderson "had made the greatest mistake of [his] life" in not "accidentally having that scoundrel killed."[33]

In characteristic fashion, Roosevelt put his anger to work, storming Capitol Hill. In describing his ongoing lobbying activities to Captain Anderson, Roosevelt promised "to use the recent unfortunate slaughter of buffaloes for all it is worth for trying to get legislation through. I haven't the least idea whether we will be successful or not. I have seen half a dozen Senators about it already." Roosevelt would also provide formal Senate testimony in support of the enforcement legislation.[34]

Captain Anderson, closely attuned to the politics of the park, had also been quick to grasp the political value of the Howell poaching episode. In an article he later wrote for a Boone and Crockett book edited by Roosevelt and Grinnell, the captain wrote that Howell's crime "has been of more service to the Park than any other event in its history."[35]

On the day that Burgess delivered Howell to Mammoth, Anderson wrote to Secretary of the Interior Hoke Smith to urge that the incident "be made the occasion for a *direct* appeal to Congress." Secretary Smith, a railroad opponent and a strong advocate of enforcement legislation, appears to have staged the release of Howell to effect the maximum political theater. In his April 2 letter ordering Howell's release, the secretary praised the Yellowstone cavalry's "zeal" in capturing the poacher. Then he reiterated what Captain Anderson knew all too well. "I have to state, that there is no law authorizing the punishment of trespassers in the Park, or persons depredating upon the game therein."[36]

Howell's release—as Anderson and Smith surely intended—only added to the growing public outrage over the killing of the buffalo, a fact reflected in congressional debate of the issue. In the Senate floor debate, Senator Vest reminded his colleagues that "all that could be done" with the notorious Howell was "to take away his old gun and a pair of blankets and put him outside the park."[37]

At *Forest and Stream*, meanwhile, Grinnell continued to stoke the fire. His most powerful argument was a simple one: Time had nearly run out. In an April 14 editorial titled "Save the Park Buffalo," Grinnell publicized the latest estimates of the number of buffalo alive in Yellowstone. Emerson Hough's expedition in the park—together with the patrolling by Captain Anderson and his men—led to the grim conclusion that the situation for the buffalo was "infinitely worse than anyone had supposed." A year earlier, Anderson had estimated the Yellowstone herd at

500 animals. But on the basis of the most recent count, Anderson and Hough now placed the total population at 200, with 250 as the "outside limit." "There seems now to be little doubt," wrote Grinnell, "that within the last year or two a wholesale slaughter has been taking place among our buffalo preserved in the Yellowstone Park." Another Grinnell editorial concluded with the blunt assessment, "At this rate it will not be long before the last shall have been shot down."[38]

Congressman Lacey's Committee on Public Lands would come to the same conclusion in urging passage of the park protection legislation: "Out of the vast herds of millions of buffaloes that a few years ago coursed the plains of America a few hundred only remain, and they are now all in the Yellowstone Park. . . . A few days ago poachers entered the park and commenced the slaughter of these animals. Prompt action is necessary or this last remaining herd of buffalo will be destroyed."[39]

By 1894, *Forest and Stream* was not the only newspaper urging protection of the buffalo. As Emerson Hough noted, perhaps a bit territorial over the story he broke, the Howell episode "has appeared in various forms and in some inaccurate shapes in the press all over the country."

In that indefinable realm where press attention and public concern coalesce into political action, the tipping point, at last, had been reached.

On April 23, 1894, the Senate passed its version of the Lacey Bill. A House–Senate conference committee (including both Congressman Lacey and Senator Vest) quickly reconciled the bills, even adding additional enforcement measures. Final passage went off without stumble, and on May 7, 1894, President Grover Cleveland signed the bill into law. Only fifty-six days had passed since the capture of Howell in a frozen Yellowstone valley. Yet this final surge rested on a foundation of two decades of relentless effort by Grinnell and his allies. Now, at last, they had prevailed.

The National Park Protective Act, known popularly as the Lacey Act, created the enforcement regime for which Grinnell had battled for so long. The act repeated the long-standing prohibition against "all hunting, or the killing, wounding, or capturing at any time of any bird or wild animal." Now though, there would be teeth behind the law. Any person found guilty would be subject to a fine of up to $1,000 and prison term of up to two years behind bars. To assist the understaffed army in its enforcement efforts, the Lacey Act provided for two new deputy marshals—with the intent that these two positions would provide the additional civilian scouts that Captain Anderson and his predecessors had long sought. To make it easier to secure convictions, the act made possession of dead animals "*prima facie* evidence" of guilt (foreclosing the common claim of poachers that game had been shot outside of park borders). To try the accused expeditiously, the act dispatched a new federal commissioner to reside permanently in Yellowstone. In a potent symbol of the new regime, the act even funded the construction of a jail within park boundaries.[40]

In a Boone and Crockett book titled *Hunting in Many Lands*, editors Grinnell and Roosevelt called the "final triumph" of passing the Lacey Act "a matter of congratulation to every sportsman interested in the protection of game, and fulfills one of the great objects sought to be attained by the foundation of the Club."[41] The act would soon be extended to Yosemite, Sequoia, and General Grant national parks. To this day, the act forms the foundation for law enforcement in all national parks.[42]

In Yellowstone today, the year 1872 is emblazoned on everything from bronze plaques to souvenir sweatshirts. Certainly it is appropriate to acknowledge the date of establishment for the world's first national park. In a real sense, however, it is the watershed year of 1894 that is far more significant. The battle to save Yellowstone and its buffalo marked the first national showdown over the environment. In the defeat of the railroad's ambitions for the park and in the passage of tough legislation to protect the buffalo, the nation—for the first time in its history—made the conscious decision to protect wildlife and wild places when it cost something, indeed when powerful commercial interests stood in direct opposition. The year 1872 marked the birth of Yellowstone as a place, but 1894 marked the birth of Yellowstone as a commitment.

No man had been more important to forging that commitment than George Bird Grinnell. In Grinnell's West, the vast land beyond the Mississippi had become, for the first time, more than the sum of its parts, more than a mere repository of raw materials to be dug up or shot down. In Grinnell's West, wildlife and wild places would be protected and revered for their own intrinsic value.

CHAPTER FIFTEEN

"Simple Majesty"

Few have done so much as you. None has done more to
preserve vast areas of picturesque wilderness for the eyes of
posterity in the simple majesty in which you and your
fellow pioneers first beheld them.

—President Calvin Coolidge
honoring George Bird Grinnell,
1925[1]

While the significance of the 1894 National Park Protective Act for the West and for the nation was profound, its significance for the buffalo would remain an open question in the decade to follow. Congressional action—if not too little—appeared to have come too late.

Reduced by the mid-1890s to 200 scattered animals, the Yellowstone buffalo had spiraled into a descent from which recovery appeared increasingly unlikely. In his darker, private moments, Grinnell doubted they could be saved. Responding to a grim 1896 report from Captain Anderson on the ever-diminishing numbers, Grinnell wrote back to express his fear that "buffalo in the United States are doomed." Natural attrition and the effects of inbreeding whittled away at the Yellowstone herd. Poaching, though greatly reduced, continued to exact a toll. The National Park Protective Act finally had given Captain Anderson the power to punish the men he caught, and his cavalrymen continued their zealous efforts. Still, the captain's correspondence from 1895 to 1897 (the end of his Yellowstone tour of duty) contains dozens of tips from informants about ongoing poaching operations.[2]

By 1902, the army estimated that only 23 buffalo remained alive in Yellowstone National Park.[3]

Yet the buffalo was not yet doomed. While Grinnell stood center stage, the war to save the buffalo was waged by many others as well. In its final chapter, the supporting cast was drawn from the diverse characters that only history can concoct, including a reformed hide hunter, a "philandering Pend d'Oreille," two "half breed" ranchers, a Texas Ranger, the former taxidermist at the Smithsonian Institution, and the president of the United States.[4]

Charles "Buffalo" Jones: While his legacy is mixed, Buffalo Jones played an important role in the conservation of wild buffalo. *Frederic Remington drew this sketch for* Harper's *magazine.*

IN 1902, THE DEPARTMENT OF THE INTERIOR HIRED A FORMER BUFFALO hunter named Charles Jesse Jones as the new Yellowstone National Park game warden, his mission to save the tattered remnant of the Yellowstone herd. In his earlier life, Jones had won fame and lost several fortunes in a variety of frontier schemes. His celebrity in 1890 was sufficient for no less than Frederick Remington to draw his portrait. In his later life he would pin his hopes for financial bonanza on inventions including a "balloon-propelling device" and a "patented water elevator." It was the bison, though, that gave him both his greatest fame and his nickname—Buffalo Jones. Jones believed that the ultimate high-plains ranch animal could be bred if only the best attributes of cattle and buffalo could be combined. He spent decades in a largely futile effort to breed what he called a "catalo."

Jones was not the only rancher to experiment with the interbreeding of cattle and buffalo. No commercially viable hybrid was ever produced, but the existence of small bands of domesticated buffalo would be important to the ultimate survival of the species. It came, though, at a cost. To this day, most living buffalo (not including the Yellowstone herd) show the genetic taint of past interbreeding with cows.[5]

While part of Jones's interest in the buffalo was commercial, he also appears to have cared deeply about the fate of the species and to have deeply regretted his former life as a commercial hunter. "I am positive it was the wickedness committed in killing so many, that impelled me to take measures for perpetuating the race which I had helped to almost destroy."[6] By the end of the buffalo hunting era in Texas, Jones had thrown aside his rifle in favor of a lariat, capturing live buffalo calves that would form the kernel of a domesticated herd. When Congress determined in 1902 that the only way to save the Yellowstone buffalo was to import pure-blooded, domesticated animals, Jones had a national reputation as a buffalo expert—including the support of President Theodore Roosevelt and George Bird Grinnell. Congress appropriated $15,000 to buy and support a domesticated herd in the park, and Buffalo Jones was hired as the Yellowstone game warden.

Jones's first task on his arrival, in the summer of 1902, was the construction of a gigantic corral to hold the domesticated herd. He chose a large field outside of Mammoth for the site. Though not part of the buffalo's traditional Yellowstone range, the site would allow close supervision by park officials. Perhaps even more significant, a location near Mammoth, then the primary entrance to the park, would ensure that tourists could see the new herd.

By October 1902, construction of the buffalo corral was complete. Now Jones needed pure-blooded buffalo (i.e., no cattle blood) to stock the herd. He acquired twenty-one animals, of which eighteen can be credited in no small part to a wayward Indian.

Three decades earlier, in 1872, a Pend d'Oreille named Samuel Walking Coyote went hunting buffalo near the Montana–Canada border. As often happened, calves orphaned or separated during the hunt began to follow the horses of the hunters. Eight such calves apparently fell in tow behind the horse of Walking Coyote.[7] At the time of his hunt, Walking Coyote was a man with a significant moral quandary. Though a Pend d'Oreille, he made his home with the Salish tribe near the Catholic mission in the valley of Flathead Lake. As Walking Coyote hunted up north, his wife, a Salish and a Catholic, awaited his return. While hunting, Walking Coyote had been living with the Blackfeet, and while living with the Blackfeet, he had taken a second wife. Walking Coyote knew that neither the Salish nor the priests who ran the mission (not to mention his Salish wife) would approve of his newest spouse, leaving him vexed as to how to return home in the spring.

The eight buffalo calves seemed to present a potential solution. Walking Coyote decided to offer them as a gift to the mission priests, hopeful that such a grand gesture might result in his absolution. The ploy culminated poorly. Two of the calves died on the voyage to the Flathead. The six surviving buffalo failed to impress either the Catholics or the Salish, perhaps because Walking Coyote also brought along his Blackfoot wife. The priests apparently ordered a beating by Indian police for both Walking Coyote and his new bride, after which the Pend d'Oreille was banished from the tribe.

Walking Coyote held on to the buffalo, and by 1884, his little herd had grown to thirteen. Having failed in their intended purpose, the buffalo had become more of a burden than Walking Coyote deemed worthwhile. Of course, they had also become a rarity. By the early 1880s, the decimation of the northern herd was complete.

Fortunately for the future of the buffalo species, Walking Coyote and two local ranchers recognized the potential value of buffalo as something living rather than dead. One of the ranchers was Charles Allard, son of an Indian mother and a French-Canadian father. The other was Michel Pablo, whose mother was Indian and father Mexican. When Allard learned that Walking Coyote might be interested in selling his herd, he arranged a meeting to discuss a price. Allard asked Michel Pablo to attend as an interpreter, and Pablo became so intrigued with the enterprise that

Many of the buffalo alive in the wild today
can trace their lineage to the herd owned
by this man, Michel Pablo.
Courtesy of the Montana Historical Society.

he asked Allard to take him on as a partner. Allard and Pablo bought
Walking Coyote's thirteen animals for $2,000. (Walking Coyote took this
small fortune to nearby Missoula, went on a bender, and exited history a
short time later, discovered dead beneath the Higgins Street Bridge.)

In 1893, Allard and Pablo added 26 more pure-blooded buffalo to
their herd, purchasing the animals from none other than Buffalo Jones.
Sheltered beneath the soaring Mission Mountains in the lush Flathead
Valley, the Pablo–Allard herd by 1896 had grown to 300. Allard died
that year, and his half of the herd passed to relatives. Some of these pure-
blooded animals ended up in Medora, North Dakota, and it was from
the Medora remnant of the Pablo–Allard buffalo that Jones obtained 18
charter members of Yellowstone's new domesticated herd.[8]

Understanding the importance of diversification among the breeding stock, Jones added at least three other animals to the Pablo–Allard
stock. As the Texas buffalo fell to hide hunters in the early 1870s, a ranch
wife named Mary Goodnight had urged her husband—Charles Goodnight, a former Texas Ranger—to preserve a few on their ranch. They
had plenty of room. In classic Texas fashion, the Goodnight property
encompassed 1 million panhandle acres. Like Buffalo Jones, Goodnight

used some of his buffalo to pursue the illusive catalo. But he also kept some of his herd pure. Jones bought three of Goodnight's pure-blooded buffalo bulls and drove them overland to the park, adding Texas blood to the Yellowstone herd.[9]

As he pursued the survival of the pure-blooded buffalo, Jones had more success with animals than with people. Just about everybody in Yellowstone, it seemed, despised him. A park concessionaire described the game warden as "pompous, opinionated, a thoroughgoing four-flusher." Jones, a teetotaler who also frowned on foul language and gambling, alienated the soldiers and scouts by tattling constantly about their behavior to the park superintendent, Major John Pitcher.[10]

As for Major Pitcher, he clashed repeatedly with Jones over a number of the game warden's tactics for handling park wildlife, including an effort to haze bears away from garbage dumps by snaring them, hoisting them upside down from trees, and beating them with a cane. President Roosevelt, who had supported Jones's appointment and who admired the domesticated herd during his famous 1903 tour of the park, apparently cut Jones out of his VIP touring group. Jones, with a tin ear to the political implications, had persistently urged the president to engage in a hunt for mountain lions. By September of 1905, the maverick-minded Jones had had his fill of government work and resigned.[11]

If Buffalo Jones had failed to make friends, he had succeeded in his primary mission. By the time he resigned, the fenced Yellowstone herd had increased from 21 to 44. By 1915, it numbered 239. And by 1916, the herd was proliferating so effectively that park officials began castrating about 50 percent of male calves.[12]

Perhaps most surprising was Yellowstone's wild herd. After bottoming out at 23 in 1902, the wild herd pawed its way back. An official report of 1912 listed the wild population at 49 and described the animals as "thriving." After 1915, the distinction between fenced and wild buffalo started to break down because park officials began encouraging the fenced buffalo to roam free. By the 1930s, Yellowstone administrators began a formal policy transition from managing the Yellowstone herd according to ranching techniques in favor of preserving the herd in a wild state. Barring unforeseen disaster, the survival of the pure-blood buffalo had been assured.

In the early twentieth century, most Yellowstone buffalo were fenced
in order to manage and to protect the herd.
Courtesy of the National Park Service, Yellowstone National Park.

That the wild buffalo survives today can be attributed, in large measure, to the Yellowstone refuge—the "single rock" that Grinnell helped
to define and then to defend. Equally significant, Grinnell's three decades
of advocacy for the environment had done much to create a nationwide
ethic—the notion, so discordant for its time, that the survival of the buffalo mattered.

While Grinnell is unique, there are many who share credit in the long
battle to save the buffalo. In 1906, Michel Pablo learned that the Flathead
Indian Reservation would be opened for homesteading. The reservation
was where Pablo grazed his 500-strong herd of buffalo, descendents of
the Walking Coyote calves. Realizing that a rush of homesteaders would
take away the open range on which his herd relied, Pablo first asked the

U.S. government for a land grant on which he could run the herd. When no offer was forthcoming, Pablo offered to sell the herd to the government. President Roosevelt embraced the offer, but Congress dragged its feet. In frustration, Pablo accepted an offer from the government of Canada.

Like the U.S. decision, the Canadian decision to protect the buffalo had almost come too late. By the time effective enforcement of protective laws was in place, the Canadian herd had dwindled to a few dozen scattered animals. With the Pablo herd, however, the Canadians would reestablish the buffalo. Pablo shipped 704 animals by rail after a Montana roundup that stretched across six years. Most of the buffalo ended up at the enormous Wood Buffalo National Park in northern Alberta, established in 1922 to provide habitat for a wild, free-ranging herd.[13]

It is worth noting that left to the U.S. government alone, the American buffalo would almost certainly have perished. Washington, it is true, set aside Yellowstone—the vital, public land that became the "single rock." But it was the relentless drive of private citizens such as Grinnell and private organizations such as Boone and Crockett that ultimately pushed the government to protect the park and its wildlife. Critical too was the vision of individuals such as Buffalo Jones, Michel Pablo, Charles Allard, and Mary and Charles Goodnight, who created private refuges within which the living seed of future generations of buffalo was preserved.

One other private group also deserves its fair share of credit. In December 1905, a meeting was held in the Lion House of the New York Zoological Park. Attendees were spare in numbers but abundant in energy. The group they established that day would be known as the American Bison Society. As president, they chose William Hornaday—the Smithsonian taxidermist who in 1886 had shot and stuffed some of the last specimens of Montana's wild herd.

By 1905, however, Hornaday's interest had evolved. Instead of seeking to present dead buffalo realistically, Hornaday began a nationwide effort to preserve the living buffalo in the wild. With Hornaday as president and Theodore Roosevelt as honorary president, the American Bison Society launched a flurry of activities that helped to ensure the buffalo's survival.[14]

The society's most important achievement was creation of the

National Bison Range on land formerly part of the Flathead Indian Reservation. West of the Continental Divide, the area was probably not part of the buffalo's historic range. But it was the same land where Walking Coyote and then Michel Pablo had raised their herds. Sadly, its availability came about only after Congress passed the Dawes Act of 1887, which converted collectively held tribal lands to homesteads. After distributing some homesteads to eligible Indians, "surplus" land was sold to the public at large.

The 13,000 acres that became the National Bison Range had marginal agricultural value. Congress, goaded by Roosevelt and embarrassed at the loss of the Pablo herd to the Canadians, appropriated $30,000 to purchase the land. Hornaday and the American Bison Society raised $10,000 to stock the reserve with its first thirty-four buffalo—descendants of the Pablo–Allard herd. Three others were donated, including one from Charles Goodnight.[15]

The American Bison Society—together with land conservation policies pioneered by Roosevelt—would help to create a number of other public buffalo reserves, including Wind Cave National Park in South Dakota, Fort Niobrara in Nebraska, and Wichita Forest Reserve in Oklahoma. In part because of Hornaday's commitment, Americans today can view buffalo beyond museum dioramas and the confined spaces of zoos. Even more significant, geographic dispersal helps to decrease the collective risk of disease, a critical element to ensuring the survival of the wild buffalo for the millennia yet to come.

By 1894, George Bird Grinnell had lived a life that embodied the adventures of a titanic age. In a West still wild, Grinnell had hunted dinosaur bones, raced prairie fires on horseback, run trap lines with mountain men, and hunted buffalo with naked Crow Indians. He had traded gunfire with the Cheyenne, traipsed through the Black Hills with Custer, and been among the earliest to gaze on the wonders of Yellowstone National Park.

Yet if Grinnell were a part of his times, there is a far greater degree to which Grinnell stood apart from his times. What is remarkable in the story of the buffalo is not that they were slaughtered, for naked exploitation was the central organizing principle of the era. What is remarkable is that amidst that era, Grinnell emerged with a vision so startlingly different. More amazing still was his ability to create from that vision a different world.

Grinnell's was a providential life, a singular combination of ethics, experiences, and skills that so thoroughly distinguished him from his times. From Lucy Audubon, Grinnell learned the small lessons in the beauty of a red crossbill and the large ones in a life committed to the welfare of others. From Professor Othniel Marsh, Grinnell learned firsthand the fragility of life—unearthing a past in which wooly mammoths and dinosaurs walked the American plains. Later, as one of the earliest visitors to Yellowstone National Park, Grinnell would be among the first to witness the emerging threat of commercial hunting to its ancient inhabitants. There were political and military lessons as well. From George Armstrong Custer, Grinnell learned the inexorable power of a tide propelled by the twin forces of commercial and strategic interest.

If the lessons of his youth were unique, Grinnell spent his adult life in a near missionary effort to make them universal. Through journalism, Grinnell bridged the gap between the abstract knowledge of academia and the practical knowledge that shapes world events. He learned the power of the pen, and *Forest and Stream* helped him to mobilize sportsmen as the nation's first national constituency for wildlife. He broadened his constituency through creation of the Audubon Society—an ode to his boyhood tutor as much as her artist husband. When railroad and real estate lobbies proved more powerful than public opinion, Grinnell created a lobby. Propelled by Roosevelt and Grinnell, the Boone and Crockett Club became the first national lobby to fight for conservation. Boone and Crockett helped to win the battle for the soul of Yellowstone Park, for a new West in which wild places and wildlife would be revered. And on the single rock of Yellowstone would the buffalo be saved.

All of that before the age of forty-five.

George Bird Grinnell, who wrote prolifically throughout his life, penned what is arguably his greatest essay while staring down at two relics of his past. Two buffalo skulls, one a bull and one a cow, lay like sentinels at either end of his hearth. In "The Last of the Buffalo," an 1892 article for *Scribner's Magazine*, Grinnell recalled how he had gathered up the skulls from the prairie, "to keep as mementoes of the past, to dream over, and in such reverie to see again the swelling host which yesterday covered the plains, and today are but a dream."[16]

For many whose lives mix with history's epochal sweep, the force of nostalgia becomes, in later life, an anchor. By his mid-forties, Grinnell might easily have fallen into the trap of those whose greatest adventures and successes reside in their past. Yet for Grinnell, his later years were marked by a frenetic activity that made it appear as if he were well aware of this pitfall, focusing furiously on the future lest he fall too often into "reverie" for his past.

Grinnell's advocacy for the buffalo and other wildlife would continue throughout his life, including an ongoing association with Boone and Crockett. From 1918 to 1927 he served as club president. In 1911, after a 35-year run, Grinnell finally gave up both the ownership and the editorship of *Forest and Stream*.[17] The same year, apparently determined not to miss a beat, he cofounded the American Game Association.[18]

Given the scale of Grinnell's life—and the contemporary relevance of his times—it is remarkable that he is not better known today. To the extent that he is known, it is often for an aspect of his career that this book has barely scratched—extensive work as an Indian ethnologist. In 1889, Grinnell published *Pawnee Hero Stories and Folk-Tales*, the first of what would grow into a library of 29 books. As he watched the closing chapters of the Old West, Grinnell was determined to preserve the history and culture of the American Indians. In research that focused on the Pawnee, the Blackfeet, and the Cheyenne, Grinnell interviewed survivors of the old days to record aspects of native life ranging from hunting and war to marriage and government. Through Grinnell, much was preserved that might otherwise have faded into the murk of time. In recognition of his efforts, the Blackfeet made Grinnell an honorary member of their tribe. Many of Grinnell's works, including classic studies of the Cheyenne Indians, are still respected (and in print) to this day.[19]

Though Grinnell and his wife would have no children of their own, he doted over a number of nieces and nephews. To amuse them (and with a tip of the hat to Mayne Reid) Grinnell published a series of adventure books for boys—the "Jack books." They were short works of fiction with titles such as *Jack in the Rockies* and *Jack the Young Explorer*, but they drew deeply from the well of his real-life adventures in the West. It is noteworthy that in 1938, as Grinnell lay bedridden in the last months of his life, he often asked his secretary to read aloud from this series.

In 1925, President Calvin Coolidge presented Grinnell with the Roosevelt Memorial Association medal for distinguished service. "Few have done so much as you," said the president at the White House ceremony. "None has done more to preserve the vast areas of picturesque wilderness for the eyes of posterity in the simple majesty in which you and your fellow pioneers first beheld them."[20]

In the same year that he received the Roosevelt medal, Grinnell succeeded Herbert Hoover as the president of the National Parks Association. It was an appropriate position for the man who fought the defining battle to preserve Yellowstone. It was another great park, though, to which history would bind an even tighter connection to Grinnell.

The last great cause of Grinnell's life was born of his passion for exploration. In the summer of 1885, Grinnell read an article submitted to *Forest and Stream* by James Willard Schultz. Titled "To the Chief Mountain," it detailed Schultz's adventures in the remote St. Mary's Lake region of the Montana Rockies, just south of the international border. To Grinnell, it must have felt as if the words themselves reached up from the page to grab him by the collar. On a continent by then mostly explored, here lay a piece of true terra incognita. Shultz described crystalline lakes "walled in by stupendous mountains," enormous trout, hunting for bighorn sheep and the illusive mountain goat. Perhaps most intriguing of all, Schultz described the capstone atop one of the soaring peaks—"a true glacier ... at least 300 feet thick." Within a matter of weeks, Grinnell was on a train headed west.[21]

The railroad reached as far as Helena. From the capital, Grinnell would travel an additional 300 miles by buckboard and then horseback. However arduous, the journey took him to the place he wanted to be—where no path marked the way. From the top of a ridgeline through morning mist, Grinnell described a vista that would inspire generations, the "stern black faces of tremendous escarpments which rose from the water's very edge." It was a brutal riot of water and ice, earth and sky— and it inspired Grinnell as no place ever had. The glaciers of St. Mary's, he told a friend, "beat anything in the whole country."[22]

In the years that followed, Grinnell returned to St. Mary's country as often as the remoteness of the region would allow. For a while he continued to carry his rifle, but as the years passed, it was other tools—the compass and the barometer—that began to take the place of his gun. He used the barometer to measure the height of the peaks he climbed, christening many with the names they bear today. Grinnell became obsessed with the drawing of a map of this last untainted place, as if to record its every detail was to savor the marrow from the last true wilderness experience of his life.

In 1892, even as he fought the uphill battle to protect Yellowstone, Grinnell first mentioned another big idea. In a letter to *Century Magazine*, Grinnell pitched an article about the St. Mary's region. The title he proposed was "The Crown of the Continent," and the article would describe the stunning physical features of the land. "[S]ome day," he added, "I hope it may be set aside as a national park."[23]

Grinnell would devote a significant portion of the next two decades to seeing this vision realized. His tools were well honed: educational articles about the richness of the resource, fiery *Forest and Stream* editorials demanding protection, Boone and Crockett lobbying, coalition-building with the citizenry of Montana, and alliance with influential members of Congress.

At several junctures it appeared that the wonders of the St. Mary's region would fall victim to a vision of a different sort. Beginning in the mid-1890s, prospectors began seeping into the region's hidden slopes and vales. By 1898, with the Klondike gold rush in full fever, the seep turned to flood. Montana's new state flag carried the motto "*oro y plata*," and for some, there were no treasures in the Treasure State but those that glittered in silver, copper, and gold.

Grinnell was skeptical about the prospects for finding mineral wealth around St. Mary's. "I have been running around over those mountains for the last eight or ten years, and never saw any great sign of mineral." Still, after his experience in the Black Hills with Custer, Grinnell understood intuitively the power of a gold rush. Boosters predicted that the St. Mary's region would "make a bigger camp than Butte." If prospectors did turn up minerals in paying quantities, Grinnell knew that his dream of a new national park could never be realized. "Paradise," he worried in a letter to a friend, "will be invaded before very long by mines."[24]

It took a decade for miners to flush the last hope for gleaming metals from the St. Mary's region. Even then, another boom would arise after oil was discovered along Swiftcurrent Creek. When Grinnell visited the area in 1904, he found everyone with "a pocketful" of oil stock. But "I believe this oil excitement will pass as the copper passed."[25]

Many more years would pass, but Grinnell's prediction would be borne out. On May 11, 1910, President William Howard Taft put his pen to the legislation that Grinnell had so long championed. The Glacier National Park Bill had become law.

ON JULY 14, 1929, 79-YEAR-OLD GEORGE BIRD GRINNELL SUFFERED a heart attack that nearly killed him. He would never fully recover, spending most of the last decade of his life in his New York City apartment. A friend encouraged Grinnell to make one last trip to Glacier National Park, attempting to entice the old man with the promise that he could "go in an auto, sleep in a feather bed, and live in a Steam Heated Room on most of our old campgrounds." Grinnell declined. That was not the West he loved.

Grinnell died on April 11, 1938, at the age of 88. The *New York Times* obituary called him the "father of American conservation." For the *New York Herald Tribune*, the death of Grinnell "cut a strong strand in the remnants of the thinning cable that still links America with the age of its frontier."[26]

He would have appreciated the praise, no doubt. But a more personal tribute, and one that would have made him particularly happy, came in a telegram from E. T. Scoyen, superintendent of Glacier National Park, to Grinnell's wife Elizabeth. Of her husband, Scoyen wrote that "he more than any person deserves the distinction of being the father of this great Park." Glacier, Scoyen believed, would always "be a monument to his memory."[27]

In Glacier Park today the visitor can hike the alpine splendor of the Grinnell Glacier Trail, peer up at Mount Grinnell, and touch the great sheet of prehistoric ice that inspired Grinnell to preserve it all for generations yet unknown.

"My glacier": An elderly George Bird Grinnell on Grinnell Glacier in his last great legacy, Glacier National Park.
Courtesy of Archives and Special Collections, University of Montana-Missoula.

THE LAST STAND

"Something Unprecedented"

If we do this right, we can create something unprecedented.

—SEAN GERRITY,
American Prairie Foundation[1]

O n May 1, 2006, a buffalo cow dropped her calf on the sparse plains of northeast Montana. Within minutes, the newborn bull had raised itself on quaking legs, its reddish hue stark against the green leaf of greasewood in spring. The birth was an insignificant event, at least as measured by the sweep of time. Over tens of thousands of years, hundreds of thousands of buffalo had been born in this low valley along a shallow creek. Yet this birth was unique: For more than a hundred years, no buffalo had been born in this place. And if not yet exactly wild, the newborn calf is genetically pure[2]—with lineage to Wind Cave National Park and, ninety years earlier, to Yellowstone.[3]

The calf is unique in another way; indeed, it is historic. It is part of a path-breaking effort by a Montana-based group called the American Prairie Foundation—in partnership with the World Wildlife Fund—to reestablish wild buffalo on their native prairie.[4]

The buffalo today is a fixture in American history and lore. We emblazon its image in our most iconic displays—state flags, official seals, commemorative coins—proud to parade this mighty animal as a symbol of ourselves. Less consciously, but even more profound, we embrace the buffalo as a metaphor for a wildness and freedom in our past that remains vital to any understanding of our national character today.[5]

Renaissance calf: Born May 1, 2006, this bull calf is the first genetically pure buffalo to be born on the Montana plains in more than a century. *Courtesy of Bill Willcut and the American Prairie Foundation.*

It is a mark of national pride that the wild buffalo has moved away from the precipice of extirpation. The Yellowstone herd is nearly 4,000 strong—all wild—including 500 calves born in the summer of 2006. The numbers are significant on other buffalo refuges as well: 400 at the National Bison Reserve, 300 at Wind Cave, 500 at the Wichita Mountains Wildlife Refuge. In Canada, more than 10,000 wild bison now populate the Wood Buffalo National Park and other public lands. In Canada and the United States, hundreds of thousands of buffalo (most of them of mixed blood) are raised as part of domesticated herds.[6]

Today it seems remarkable that the buffalo once stood so close to extinction. That the animal should have been saved seems self-evident. History, somehow, lends clarity that is difficult to discern when the setting is more proximate in time. Grinnell's genius was his ability to see the future before it was too late, and then to act, and then to act with great effect.

The threats to the environment today are more subtle than the crack of the Sharps buffalo rifle. Yet with each generation our human footprint has grown larger, until what is threatened is not just individual species but also whole ecosystems. The buffalo has been saved from extinction, but today the prairie itself is endangered; less than 2 percent of America's native prairie is protected.[7]

Our modern society may not think in terms of conquering the wilderness, yet our domination of the environment is so complete that even the weather is now shaped by human activity. In Glacier National Park today, visitors can still see the bighorn sheep and mountain goats that thrilled Grinnell. But by 2030, Grinnell Glacier will be gone. Gone, like every other glacier in Glacier National Park, a victim of global warming.

The nineteenth-century activism of the Grinnell generation, with its zenith in the presidency of Theodore Roosevelt, created the government protection model that defined conservation in the twentieth century. In the most intriguing conservation model of the twenty-first century, groups such as the American Prairie Foundation and the World Wildlife Fund aren't waiting for government to act. On the plains of northeast Montana, using private funds to purchase private property, the American Prairie Foundation and the World Wildlife Fund have pieced together 32,000 acres on which they intend to run buffalo. Their ultimate vision is breathtaking in its scope, with hopes of one day establishing a reserve that encompasses 3 million acres. If successful, this "prairie project" would be half again as large as Yellowstone National Park.[8]

"If we do this right," says Sean Gerrity, the Montana-born president of the American Prairie Foundation, "we can create something unprecedented."[9] What the prairie project aims to do goes far beyond the buffalo, though a free-ranging herd of genetically pure bison would be a signature achievement. The broader goal is a "prairie-based wildlife reserve." Such a reserve would also provide habitat for the prairie dog, burrowing owls, mountain plovers, and the black-footed ferret—the most endangered mammal in North America. A fragile prairie plant ecology would also have the opportunity to thrive. Indeed, one of the reasons northeastern Montana was selected for the prairie project is that the land there is remarkably intact, owing in part to the twentieth century's stewardship by local ranchers.

Echoing Grinnell, a foundational tenet of the prairie project philosophy is "public access." Though privately organized, its central aim is to prevent the land from being "locked up for the benefit of a select few." Reflecting the Western values of its location—and building on Grinnell's coalition of hunters and anglers—the preserve will one day welcome hunters. In a part of the country that reveres private property rights, the American Prairie Foundation and the World Wildlife Fund are defusing some critics through their "willing buyer, willing seller" model of acquiring lands. Some local ranchers remain suspicious of the project, but it will displace no land owner who does not choose to sell.

All in all, both a vision and a model for achieving it with which Grinnell would have found much to approve.

THE HISTORY OF HUMAN INTERACTION WITH THE BUFFALO IS AT THE same time disheartening and inspiring, a gut-wrenching parable both of man's ability to do great harm—and to achieve great good.

Why was the buffalo nearly destroyed? Certainly greed played a role. So too the powerful drive to prevail in a messy war against the Indian. But more vexing, and more relevant to our own situation today, are other factors. Most of the men who pulled triggers on the buffalo weren't acting on naked greed—they just wanted to make a living. And more fatal than the Sharps was the widespread belief in the inevitability of the buffalo's destruction—that the actions of individuals could have no measurable impact against the inertial tide of human affairs.

The great lesson of George Bird Grinnell is that one person can make a difference, indeed all the difference. It's why wild buffalo walk the earth today. It's why there is still hope.

Acknowledgments

A book is a long haul, so I'm grateful for the many friends and family who have supported me along the way. Thanks to Max Baucus, Mike Bridge, Bruce Bugbee, Amy and Sean Darragh, Ken Doroshow, Allen Fetscher, Phil Gardner, Cheryl and Brent Garrett, Mickey Kantor, Jen Kaplan, Peter Lambros, David Marchick, Amy and Mike McManamen, the Miller Family, Bob Pack, Sharon Peterson, Tim and Lori Punke, Peter Scher, Fred Shellenberg, Jo May and Brian Salonen, Monte Silk, Mary and Bill Strong, Kim Tilley, Bev Whitt, Pat Williams, Lynne and Gary Willstein, and Jay Ziegler.

I am again indebted to a great many librarians and archivists for their dedication and expertise. Thanks to the staff at the University of Montana's Mansfield Library and especially to Donna McCrae in the K. Ross Toole Archives. Thanks to Harold Housley at the Yellowstone National Park Archives and to Brian Shovers at the Montana Historical Society Archives.

I have benefited greatly from the work of many scholars who have explored some of the same ground covered in this book, including a few who deserve acknowledgment beyond footnotes. John Reiger, a former head of the Connecticut Audubon Society, helped to ensure the preservation of Grinnell's papers and wrote an important book called *American Sportsmen and the Origins of Conservation*. Duane Hampton, emeritus professor at the University of Montana, gave me a big box of his research about Grinnell, valuable feedback on an early draft of *Last Stand*, and a great book (his)—*How the U.S. Cavalry Saved Our National Parks*—which captures this little-known chapter in the history of the U.S. Army. Among the many wonderful books about buffalo, I found *Head, Hides & Horns* by Larry Barsness to be a particularly comprehensive and readable resource. Thanks to Lee Whittlesey, park ranger and scholar of Yel-

lowstone, for his insights to Grinnell and the history of the park. Whittlesey, it is worth noting, is leading an effort to name a peak in Yellowstone
in Grinnell's honor – a project certainly worthy of broad support. Thanks
also to Jack Reneau at Boone and Crockett.

For sharing a glimpse of their vision of wild buffalo on the plains,
thanks to Sean Gerrity, Scott Laird, and Bill Willcutt at the American
Prairie Foundation. Anyone interested in more information about this
project should visit www.americanprairie.org or www.worldwildlife.
org. I'm also grateful to Curt Freese at the World Wildlife Fund for
lending a scientific eye to review of this manuscript.

A big thanks to T.J. Kelleher, my editor at Smithsonian Books. Beyond
wordsmithing, T.J. helped to shepherd this book with his knowledge of
a frighteningly diverse range of topics—from eighteenth-century phi-
losophers to the GI tract of a buffalo. Thanks also to Gretchen Crary
with Harper Collins.

No one could ask for a better agent than Tina Bennett at Janklow &
Nesbit. She has my eternal and heartfelt thanks. Thanks also to Tina's
assistant, Svetlana Katz, ever helpful in ways big and small.

For their ongoing support of my writing, I'm grateful to Keith Red-
mon and David Kanter at Anonymous Content.

A special thanks to my parents, Butch and Marilyn Punke, for having
the good sense to raise their children in the West. My dad took me
hunting and fishing from the time I could walk—my first exposure to
the beauty of mountains and plains. Like Grinnell, my father is part of
the long tradition of sportsmen-conservationists. He was giving slide
programs about protecting the environment to audiences in small-town
Wyoming back in 1973.

Thanks most especially to my family, who turned the research for
this book into memorable adventures. Thanks Bo, for helping to dodge
that big bull buffalo on Indian Creek, for exploring Soda Butte, and for
sharing a wet tent. Thanks Sophie, for climbing the buffalo jump in
Madison Valley, for digging through dusty documents in the Yellowstone
Archives, and for sharing a cold tent. And thanks Traci—for the VW, the
CMR, and your birthday in snowy Yellowstone.

Notes

Prologue: The Stand

1. The description of Vic Smith's hunt is drawn from Victor Grant Smith, *The Champion Buffalo Hunter* (Jeanette Prodgers, editor, 1997). The original manuscript was titled *Vic G. Smith of Montana* and was discovered (in 1990) by editor Jeanette Prodgers in the Theodore Roosevelt Collection at Harvard's Houghton Library. Though Smith wrote in the third person, the manuscript is believed to be an account of his true, first-person experiences. For the quotations in this section, I therefore took the liberty of changing Smith's use of the third person to the first-person *I*.

Chapter 1: "Wild and Wooly"

1. "We supposed they would soon pass by . . ." is quoted from Grinnell's letter to F. G. Webber (January 22, 1892) and can be found in the *Letter Book* at page 727. The *Letter Book* is an unpublished collection of Grinnell's correspondence between 1886 and 1929. The original collection is at Yale. The author viewed portions of the *Letter Book* on microfilm at the K. Ross Toole Archives, Mike and Maureen Mansfield Library, University of Montana, Missoula. "[T]hree hours" is quoted from Grinnell's letter to P. J. McGill (January 18, 1892), *Letter Book* at pages 713–714: "[W]e were twice delayed by buffalo, on one occasion having to stop three hours in order to let a herd pass." See also Grinnell's letter to E. W. Nelson (May 11, 1916), *Letter Book* at page 116.

2. George Bird Grinnell, "An Old-Time Bone Hunt," *Natural History* (July–August, 1923), at page 330.

3. Grinnell, *Memories*, at page 25. Apparently written in 1915, George Bird Grinnell's unpublished *Memories* provides some of the best insights to his early life. For reasons unknown, Grinnell stopped work on the document in the middle of a sentence, never returning to it. The original *Memories* document is at Yale. The author viwed a copy at the K. Ross Toole Archives, Mike and Maureen Mansfield Library, University of Montana, Missoula. (Note: My citations to the *Memories* document correspond to the typed, single-spaced, 62-page text on file at the University of Montana.)

4. *New York Herald Tribune*, April 12, 1938, page 14; Grinnell, *Memories*, at page 2.

5. Grinnell, *Memories*, at page 29.

6. George Bird Grinnell, *Audubon Park: The History of the Site of the Hispanic Society of America and Neighbouring Institutions* (1927), pages 1, 9, 18; Grinnell, *Memories*, at page 21.

7. Grinnell, *Memories*, pages 14, 17.

8. On Reid, see Joan Steele, *Captain Mayne Reid* (1918). Grinnell is quoted from *Memories* at page 24.

9. Grinnell, *Memories*, at page 19; George Bird Grinnell, "Recollections of Audubon Park," *The Auk* (July 1920), at page 378.

10. Grinnell, *Memories*, at page 21. See also George Bird Grinnell, "A Chapter of History and Natural History in Old New York," *Natural History*, volume XX (1920), pages 23–27.

11. Grinnell, *Memories*, at page 13; Grinnell, "Recollections of Audubon Park," page 379.

12. Grinnell, *Memories*, at page 13.

13. Lucy Audubon quoted in Francis Hobart Herrick, *Audubon the Naturalist* (1917), page 300.

14. Grinnell, "Recollections of Audubon Park," at pages 375–376.

15. The list of Churchill supplies is from an essay, "St. John's Military School: Only a Memory," at www.jahnweb.com/ajj/stjohns/history.html. Discussion of Grinnell's time at Churchill is from Grinnell, *Memories*, at page 22, and Grinnell's letter to Jesse Monroe, December 27, 1911, found in *Letter Book* at page 681.

16. Grinnell, *Memories*, at pages 22–23.

17. Ibid. at page 23.

18. "[I]n great fear . . ." is quoted from Grinnell's letter to E.S. Dana, August 21, 1908, *Letter Book* at page 702.

19. Grinnell recounts the run-up to the Marsh expedition in *Memories* at pages 24–25.

20. Biographical information on Marsh is drawn from George Bird Grinnell, "Sketch of Professor Marsh," *The Popular Science Monthly* (September 1878), pages 612–617, and Edwin H. Colbert, *The Great Dinosaur Hunters* (1968), pages 66–70.

21. The *Protohippus* anecdote and Marsh quotations are from William H. Goetzmann, *Exploration and Empire* (1966), pages x and 425.

22. Grinnell, "An Old-Time Bone Hunt," at pages 330–332.

23. "The most celebrated prairie man alive" is quoted from Grinnell's journal, July 14, 1870, a copy of which is on file at the K. Ross Toole Archives, Mansfield Library, the University of Montana, Missoula. The general description of the Marsh expedition is from C.W. Betts, "The Yale College Expedition of 1870," *Harper's New Monthly Magazine* (October 1871), page 663.

24. The description of the soldiers and Cody's quotation are from Betts at pages 663–664. The "interested auditor" quote is from Grinnell, "An Old-Time Bone Hunt," at page 332.

25. The quote "had seen quite enough of Nebraska" and descriptions of difficult water conditions are from Grinnell, "An Old-Time Bone Hunt," at pages 332–333. "110 degrees" is from Betts at page 664.

26. "To shoot at the antelope" is from Grinnell, "An Old-Time Bone Hunt," at page 333. See also Grinnell, *Memories*, at page 26; Betts at page 664.

27. Grinnell, journal, July 15, 1870.

28. Betts at page 666. See also Grinnell, journal, July 15, 1870.

29. Betts at pages 664–665.

30. Grinnell, "An Old-Time Bone Hunt," at page 333; Betts at pages 665–666.

31. Grinnell, "An Old-Time Bone Hunt," at page 334.

32. Betts at page 668.

33. Accounts of the duck hunting, the prairie fire, and the night alone on the plains are from Grinnell, *Memories*, at pages 26–27, and "An Old-Time Bone Hunt," at page 334. Grinnell fails to mention, in his own accounts, that one of the two hunters touched off the prairie fire through careless handling of a match. That fact comes from Betts at page 668.

Chapter 2: "Self-Denial"

1. "[A]t an inconceivably rapid rate" is from "An Old-Time Bone Hunt," *Natural History* (July–August 1923), at page 334; "checked the flames . . ." and "near enough to singe the hair . . ." are from Grinnell, *Memories*, at page 27. For background on the *Memories* document, see Chapter 1, note 3.

2. For general background on Vanderbilt and the Gilded Age, see Sean Dennis Cashman, *America in the Gilded Age* (1984), pages 35–36. The quotes are from Meade Minnigerode's *Certain Rich Men* (1927), at page 128.

3. Most of the background information on Lucy Audubon comes from Carolyn E. Delatte, *Lucy Audubon* (1982).

4. "*Je ne sais quoi*" is quoted from Maria R. Audubon (editor), *Audubon and His Journals* (1897), page 18.

5. One excellent resource on the life of John James Audubon is Francis Hobart Herrick, *Audubon the Naturalist: A History of His Life and Times* (1917). Another is more recent: Richard Rhodes, *John James Audubon: The Making of an American* (2004).

6. Grinnell, *Memories*, at pages 12–13.

7. Grinnell, "Recollections of Audubon Park," at page 373.

8. Ibid. at page 379. A photo of Lucy's note to Grinnell accompanies the article.

9. George Bird Grinnell, "The Character of John James Audubon," *Audubon Magazine* (October 1887), at page 196.

10. John R. Reiger, *The Passing of the Great West: Selected Papers of George Bird Grinnell* (1972), page 22.

11. Grinnell (writing as "Ornis"), "The Green River Country," *Forest and Stream* (November 13, 1873), page 212.

12. Grinnell, "An Old-Time Bone Hunt," at page 335; see also *Memories* at page 28.

13. Grinnell, *Memories*, at page 28.

14. George Bird Grinnell (writing as "Ornis"), "A Day with the Sage Grouse," *Forest and Stream*, November 6, 1873, page 159.

15. Grinnell, "The Green River Country," at page 212.

16. Ibid.

17. "[A] glimpse . . ." and "a large flock" is from Grinnell, "An Old-Time Bone Hunt," at page 335. "[T]hey still supported themselves . . ." is from Grinnell, *Memories*, at page 28. "[S]pent some days . . ." is from an unpublished 8-page document by Grinnell titled "A Memory of Fort Bridger," page 7, on file at the K. Ross Toole Archives, Mansfield Library, the University of Montana, Missoula.

18. "The river bottom and hills were full of game . . . ," "one, two or more beaver," and "I desired enormously . . ." are from Grinnell, *Memories*, at page 28. "Their mode of life . . ." is from Grinnell, "A Memory of Fort Bridger," at page 7.

19. "And then spent several weeks . . ." and "flirted with twenty-two daughters . . ." is from Betts at page 671. "Big Trees" is from Grinnell, *Memories*, at page 28.

20. "100 extinct species" is from Grinnell, "Sketch of Professor O.C. Marsh." On the scientific impact of the Marsh expedition, including Darwin's reaction, see Goetzmann at pages 425–429. According to one source, Darwin considered Marsh's discoveries so significant that he contemplated a trip to the United States "for the sole purpose of seeing Marsh's collection at Yale." See "Othniel Charles Marsh's Proof for Darwin's Theory of Evolution," http://www.geocities.com/ResearchTriangle/Lab/3773/OC_Marsh.html. The *Harper's* article, "The Yale College Expedition of 1870," appeared in the October 1871 edition at page 663.

21. Grinnell, "A Memory of Fort Bridger," at page 8.

22. Goetzmann at pages xiii, 429.

23. For a discussion of fossil hunting in the West and Grinnell's thinking on the vulnerability of nature, see John F. Reiger, *The Passing of the Great West: Selected Papers of George Bird Grinnell* (1972), page 55.

24. Hastings quoted by Harlan Hague, "Eden Ravished," *The American West* (May/June 1977), at page 33.

25. Grinnell, "Green River Country," page 212.

26. Grinnell, *Memories*, at page 29.

27. Ibid. at page 28.

Chapter 3: "Barbarism Pure and Simple"

1. George Bird Grinnell, *Memories*, at page 25. For background on the *Memories* document, see Chapter 1, note 3.

2. Ibid. at page 35.

3. Ibid. at pages 29–30.

4. "In the summer of 1872 . . ." is from *Memories* at pages 29–30. "[T]hese hunts of the Indians . . ." is from George Bird Grinnell (writing as "Ornis"), "Buffalo Hunt with the Pawnees," *Forest and Stream* (December 25, 1873), page 305.

5. Information on the evolutionary history of the modern buffalo and other mammals is from Dale F. Lott, *American Bison* (2002), at chapter 7, "Ancestors and Relatives." See also Mary Lee Guthrie, *Blue Babe* (1988), at pages 3, 15–16.

6. On the theories for the extinction of Pleistocene mammals and the advantages of *Bison bison*, see Lott at pages 64–65.

7. Lott at page 45.

8. On the range of climate and topographical conditions in which buffalo thrive, see Larry Barsness, *Head, Hides & Horns* (1985), at pages 24–25.

9. Lott at pages 48–49.

10. On buffalo birth rates, see Lott at page 19 and Barsness at page 23. On the birth rates of dairy cows, see the United States Department of Agriculture, "High-Fertility Proteins Enhance Reproduction Rates in Dairy Cattle" (October 1999), www.csrees.usda.gov/nea/animals/pdfs/oct99.pdf.

11. Lott at page 30.

12. Barsness at page 23.

13. On the buffalo's speed, endurance, and agility, see Lott at pages 41–42 and Barsness at page 26.

14. Wolves did have a significant impact on calves. According to Lott's "educated guess," wolves probably killed an average of about one-third of buffalo calves. See Lott at page 75.

15. Lott at chapter 8, "How Many?"

16. Lott at pages 73, 170.

17. General descriptions of Grinnell's buffalo hunt with the Pawnee are from two documents by George Bird Grinnell. The first, "Buffalo Hunt with the Pawnees," appeared as an article in *Forest and Stream* on December 25, 1873, at pages 305–306. The second is the chapter "A Summer Hunt" that appeared in Grinnell's book-length treatment of Pawnee life, *Pawnee Hero and Folk-Tales*, first published in 1889. In some passages the two accounts are identical. In others there are slight differences, usually nonsubstantive. The version in *The Pawnee Hero and Folk-Tales* generally offers more background information, while the earlier *Forest and Stream* version has a few more details of a contemporaneous nature.

18. Grinnell, *Pawnee Hero Stories and Folk-Tales*, at page 275.

19. Grinnell, "Buffalo Hunt with the Pawnees," at page 305.

20. Barsness at page 37.

21. Jesse D. Jennings, *Prehistory of North America* (1968), page 72.

22. On Folsom Man and his hunting techniques, see generally Jesse D. Jennings, *Prehistory of North America* (1974), pages 104–109.

23. For an extensive discussion of primitive hunting techniques, see Barsness at chapter V, "Hunting the Buffalo on Foot."

24. Lewis is quoted from the *Journals of the Lewis and Clark Expedition* (volume 4), Gary E. Moulton, editor, at pages 216–217.

25. On American Indians' earliest use of the bow and arrow, see "American Indian

Technology," Office of the State Archaeologist at the University of Iowa, www.uiowa.edu. Radisson is cited by Barsness at page 42.

26. On the Indian and the horse, see E. Douglas Branch, *The Hunting of the Buffalo* (1929), pages 22–25; Lott at pages 161–163; Barsness at chapter VI, "Hunting the Buffalo on Horseback."

27. Barsness at page 74.

28. Information and quotations concerning the buffalo hunt are from Grinnell, *Pawnee Hero Stories and Folk-Tales*, pages 270–302, and/or Grinnell, "Buffalo Hunt with the Pawnees," *Forest and Stream*, December 25, 1873, pages 305–306.

29. Grinnell, *Pawnee Hero Stories and Folk-Tales*, pages 295–296.

30. Ibid. at page 298. In Grinnell's "Buffalo Hunt with the Pawnee" version, he estimated the size of the herd at 500 (see page 306).

31. Grinnell, *Pawnee Hero Stories and Folk-Tales*, at pages 299–300.

32. Ibid. at page 300.

33. Grinnell, "Buffalo Hunt with the Pawnees," page 306. On Two-Lance's shot, see Branch at page 146.

34. Grinnell, "Buffalo Hunt with the Pawnees," page 306.

35. Sergeant Pryor quoted from Branch at pages 107–108.

36. Grinnell, *Pawnee Hero Stories and Folk-Tales*, pages 300–301.

37. On the uses of buffalo from birth to death, see, e.g., Barsness at chapter VIII, "Buffalo: The Indian's Manna."

38. Grinnell, "Hunting Buffalo with the Pawnee," at page 306.

39. Grinnell's letter to George E. Hyde, November 4, 1921, *Letter Book*, page 883. For background on the *Letter Book*, see Chapter 1, note 1.

40. Grinnell, "Buffalo Hunt with the Pawnee," at page 305.

Chapter 4: "I Felled a Mighty Bison"

1. William Marshall Anderson, "Anderson's Narrative of a Ride to the Rocky Mountains in 1834," *Frontier and Midland* (edited by Albert J. Partoll, Autumn 1938), at page 55.

2. Solis cited by Ernest Thompson Seton, "The American Bison or Buffalo," *Scribner's Magazine* (October 1906), at page 385. See also E. Douglas Branch, *The Hunting of the Buffalo* (1929), pages 11–12. One historian, Francis Haines, casts some doubt on the de Solis account, claiming that it did not appear in any of the original accounts. See Haines, *The Buffalo* (1970), pages 28–29. One possible explanation, not discussed by Haines, is that de Solis, writing in seventeenth-century Spain, had access to documents and sources not available in later times.

3. "The Narrative of Alvar Nuñez Cabeça de Vaca," (edited by Frederick W. Hodge), *Spanish Explorers in the Southern United States*, pages 1528–1543 (1907), pages 68, 99–105.

4. Seton at page 385.

5. On buffalo in the eastern United States, see Haines at chapter 9, "Buffalo East of the Mississippi."

6. Emphasis added. The full text of the letter, dated June 20, 1803, can be found in Donald Jackson (editor), *Letters of the Lewis and Clark Expedition* (1978), 1:61–66.

7. Lewis and Clark journal citations are from *The Journals of the Lewis & Clark Expedition* (2001), Gary E. Moulten, (editor).

8. Lewis's report is quoted by David Lavender, *The Way to the Western Sea: Lewis and Clark Across the Continent* (1988), page 374. See also Stephen E. Ambrose, *Undaunted Courage: Meriwether Lewis, Thomas Jefferson, and the Opening of the American West* (1996), at pages 396–398.

9. Information on the beaver and the Hudson's Bay Company is from Parks Canada at www.pc.gc.ca.

10. On teamsters, see Larry Barsness, *Heads, Hides, & Horns* (1985), at page 112.

11. Branch at page 100.

12. On numbers of buffalo robes traded, see Theodore R. Davis, "The Buffalo Range," *Harper's New Monthly Magazine* (January 1869), page 163, and Branch at pages 103 and 125. For extracts of the Chouteau–Crooks correspondence, see Branch at pages 95–102.

13. Audubon journal entry quoted by Francis Hobart Herrick in *Audubon the Naturalist* (1917), volume II, at pages 255–256.

14. *Diaries and Journals of Narcissa Whitman*, June 4, 1836. Available online at www.isu.edu.

15. Ibid. at June 27, 1836.

16. "I never saw . . ." is from ibid., June 4, 1836. "We have had no bread . . ." is from Ibid., June 27, 1836.

17. Ibid., June 3, 1836. The reference to *bois de vache* is from Davis at page 153.

18. *Diaries and Journals of Narcissa Whitman*, late July or early August (exact date missing from journal).

19. On the life and death of Narcissa Whitman, see Julie Ray Jeffrey, *Converting the West: A Biography of Narcissa Whitman* (1991).

20. One of the men who helped lead the 1843 emigrants to Oregon was Marcus Whitman, returning from church business in the East. On early western emigration, see two books by David Lavender, *Westward Vision: The Story of the Oregon Trail* (1963), pages 350–351, and *The Great West* (1965), chapter VIII.

21. Anderson at page 55.

22. Stansbury and Greg quoted by Branch at pages 115–116, 119. On the "plinkers," see Barsness at chapter XI, "Plinking at the Buffalo: The Thing to Do."

23. Branch at 115.

24. Obadiah Oakly, *Expedition to Oregon* (1914), page 11.

25. Haines at page 141.

26. Lavender, *The Great West*, page 244.

27. The "35,000 miles" statistic is from "People and Events: The Panic of 1873," viewed at www.pbs.org.

28. Stephen E. Ambrose, *Nothing Like It in the World* (2000), at page 369.

29. Ferguson cited by Ambrose at page 267.

30. Russell Osborne, *Journal of a Trapper, or, Nine Years in the Rocky Mountains, 1834–1843* (1955, Aubrey L. Haines, editor), at page 138.

31. George Catlin, *North American Indians* [1989 edition edited by Peter Matthiessen; originally published as *Letters and Notes on the Manners, Customs and Conditions of the North American Indians Written During Eight Years Travel (1832–1839) Amongst the Wildest Tribes of Indians of North America*], at page 259.

32. Catlin at page 259. For other contemporaneous accounts of wasteful killing by Indians see, e.g., Lieutenant R.E.Thompson, *Report of a Reconnaissance of Judith Basin, and of a Trip from Carroll to Fort Ellis, Via Yellowstone River,* at page 56. Thompson's report is appended to a longer report by William Ludlow, *Report of a Reconnaissance to Yellowstone National Park*, 1876, viewed by the author at the Yellowstone National Park Archives in Gardiner, Montana.

33. Dale F. Lott, *American Bison* (2002), at pages 172–173.

34. On the impact of emigrants on the herd, see Davis at 152. On searching for forage, see Haines at pages 143–144.

35. Haines at pages 174–175.

36. For Cody's account of the hunt, see William F. Cody, *The Life of Buffalo Bill: Or, The Life and Adventures of William F. Cody, as Told by Himself* (originally published in 1879), at chapter 15, "Champion Buffalo Killer."

37. Davis at page 149.

38. Cited by Barsness at pages 105–106.

39. Lott devotes two chapters to analysis of both historic and prehistoric bison numbers in America (chapters 8 and 21). The quote "One may assume . . ." is from J.H. Shaw and is cited by Lott at page 168.

Chapter 5: "The Guns of Other Hunters"

1. Charles Jesse Jones, *Buffalo Jones' Forty Years of Adventure* (1899), at page 235.

2. J. Wright Mooar's story is told in a series of 1933 articles titled "Buffalo Days" in *Holland's* magazine (January through July) with the byline "as told to James Winford Hunt." His account of his earliest western experiences and the birth of the buffalo hide industry can be found in the January edition at page 13. The account of his brother joining him is from the February edition at page 10.

3. Frank Mayer with Charles B. Roth (editor), *The Buffalo Harvest* (1958), page 15.

4. Ibid. at page 16.

5. Mayer's descriptions of McRae and Vimy are found on pages 23–25.

6. W. Skelton Glenn, "The Recollections of W.S. Glenn, Buffalo Hunter," *Panhandle Plains Historical Review* (1949), at page 20.

7. Mayer at page 26.

8. On the surround and other early hunting techniques, see Barsness, *Head, Hides & Horns*, at pages 113–114.

9. Glenn at page 21.

10. Either Mayer's math or his knowledge of the president's salary was shaky. The actual salary of the president in 1873 was $50,000, or $4,167 a month.

11. "I was a businessman . . . ," Mayer at page 32. "[A]lways was that dream . . . ," Mayer at page 68. Mayer's earnings projections are at pages 61–62. His actual earnings are outlined at pages 63–64.

12. "The whole West . . ." is from Mayer at page 21. The 10,000–20,000 figure is cited by Barsness at page 113.

13. George Bird Grinnell, *Memories*, at page 33. For background on the *Memories* document, see chapter 1, note 3.

14. On the Panic of 1873, see John Lubetkin, *Jay Cooke's Gamble: The Northern Pacific Railroad, The Sioux, and the Panic of 1873* (2006), at pages 276–283.

15. Grinnell's father had actually retired (temporarily, as it would turn out), only two weeks before the onset of the Panic of 1873, passing the firm to his son under the new name, George B. Grinnell & Co. In the chaos about to ensue, Grinnell's father came out of retirement to resume his leadership role.

16. Grinnell, *Memories*, at pages 34–35.

17. On Grinnell's father's ethics, see John F. Reiger (editor), *The Passing of the Great West: Selected Papers of George Bird Grinnell* (1972), at page 7.

18. Grinnell, *Memories*, at page 34.

19. Mayer at page 15.

20. Background information on typical buffalo hunting outfits and their equipment is from John Cloud Jacobs, "The Last of the Buffalo," *The World's Work* (January 1909), pages 11,098–11,100; Mayer at pages 49–50; Harlan B. Kauffman, "Hunting the Buffalo," *Overland Monthly* (August 1915), page 166.

21. Mayer at pages 49–50.

22. "[S]ixteen hundred pounds of lead . . . ," "put back in the heat," and "It frequently happened" are from Jacobs at page 11,099. On Mayer's powder and bullet preferences, see Mayer at pages 41 and 56.

23. Barsness at page 115.

24. "[P]raying to the hunting god . . ." is from Jacobs at page 11,100; "if my life depended on one shot . . ." is from Mayer at pages 53–54.

25. "[S]mall fortune" is from Mayer at page 54.

26. "I didn't have to enquire . . ." is from ibid. at page 57; 269 buffalo with 300 cartridges is from ibid. at page 37.

27. "Shoots today, kills tomorrow" is quoted from Mooar, January 1933, at page 24. On the Battle of Adobe Walls and the 1,500-yard shot, see T.R. Fehrenbach, *Lone Star: A History of Texas and the Texans* (1968), at pages 545–546. A Comanche account of the Battle of Adobe Walls appears to describe the same shot. "The buffalo hunters were too much for us. They stood behind adobe walls. They had telescopes on their guns. . . . One of our men was knocked off his horse by a spent bullet fired at a range of about a

mile. It stunned, but did not kill him." W. S. Nye, *Carbine and Lance* (1937, 1969 edition quoted), at page 191.

28. For a detailed description of hunting techniques, see Jacobs at pages 11,099–11,100. For similar descriptions by other hunters, see Kauffman at page 166; John R. Cook, *The Border and the Buffalo* (1907), at pages 164–167; and Mayer at pages 20, 37, and 46.

29. Jacobs at page 11,099.

30. Mayer at page 20.

31. There are other claims on this title. Charlie Roth apparently shot 107 buffalo in a single stand on the Canadian River in 1873; see Branch at 200. Tom Nickson supposedly shot 120 buffalo in forty minutes; see Barsness at page 117. John Mayer claimed that Billy Dixon shot "120 without moving his sticks"; see Mayer at page 35.

32. Cook at page 87.

33. For the description of Cook's stand, see Cook at pages 164–167.

34. Mayer at page 35.

35. "Touch a skinning knife ..." is from Mayer at page 58; "dirty, disagreeable ..." is from Mayer at page 47. On skinning numbers, see Barsness at page 121.

36. On the hygiene of skinners, see Barsness at page 120.

37. Cook at pages 84–85. On Wyatt Earp, see Stuart N. Lake, "The Buffalo Hunters," *Saturday Evening Post* (October 25, 1930), pages 12–13.

38. On bloating, see Cook at 118.

39. Mayer at page 47.

40. On the hide caravans and Cook's European agent, see Cook at pages 134, 137, and 151.

41. See Mooar, May 1933, at page 12.

42. See Mooar, February 1933, at page 10, and Mooar, May 1933, at page 12.

43. "[T]he stench was so great ..." is from Mayer at page 91; "that they would look like logs ..." is from Glenn, "The Recollections of W.S. Glenn, Buffalo Hunter," at page 24.

44. Mayer at page 91.

45. Cook at page 136. On rain, see Mayer at page 62, describing how extended wet weather once caused one-quarter of his hides to molder. On less proficient crews, see Mayer's discussion of his early days on the range at page 89.

46. "[I]t was the opinion ..." is from Cook at page 125; "expected to be buffalo hunters" is from Jacobs at page 11,099.

47. Mayer at page 86.

48. "It wasn't long after I got into the game ..." is from Mayer at page 30; "Often in those bloody years ..." is from Jones at page 235.

49. "As I walked back through ..." is from Cook at page 166; "The words were sung all over the range ..." is from Cook at pages 292–293. The song to which Cook referred, of course, is "Home on the Range." Today it is the state song of Kansas.

50. Medicine Lodge Treaty (1867), at Article 2. The text of the treaty is available at http://www.nps.gov/fols/Plains_Indians/Treaty/body_treaty.html. For background on the Medicine Lodge Treaty, see Robert M. Utley, *The Indian Frontier* (1984), at pages 112–116.

51. Mooar's account of the discovery of the Texas herd and its fallout are from the February installment of his series for *Holland's* at page 44. See also J. Wright Mooar, "The First Buffalo Hunting in the Panhandle," *West Texas Historical Association Yearbook* (1930), at pages 109–111.

52. Francis Haines, *The Buffalo* (1970), at page 196. For Mooar's description of an army officer attempting to warn him off of hunting on Indian lands, see Mooar, "Frontier Experiences of J. Wright Mooar," *West Texas Historical Association Yearbook* (1928), at page 90.

53. Fehrenbach at page 545.

54. "[T]he passing of both the Indian and the buffalo . . ." is from Kauffman at page 170; "With the end of the buffalo . . ." is from Jacobs at page 11,100.

55. "Whether the ends justified the means . . ." is from Earp quoted by Stuart N. Lake, "The Buffalo Hunters," *The Saturday Evening Post* (October 25, 1930), at page 12.

Chapter 6: "That Will Mean an Indian War"

1. Grinnell's *From Notebooks of Black Hills Expedition*, entry dated July 26, 1874, is quoted from John F. Reiger, *The Passing of the Great West: Selected Papers of George Bird Grinnell* (1972), at page 100.

2. Information in the introductory section is from George Bird Grinnell, *Memories*, at pages 33–34. For background on the *Memories* document, see chapter 1, note 3. On the role of scientists in army explorations of the day, see William H. Goetzmann, *Exploration and Empire* (1966), page 391.

3. Grinnell, like most of Custer's men, referred to Custer in 1874 as "General Custer." Custer earned the brevet, or temporary rank, of brigadier general as a 23-year-old Civil War hero. After the war, Custer, like other officers with brevet ranks, was returned to peacetime status—in Custer's case, the rank of lieutenant colonel.

4. On Indian removal, see Grant Foreman, *Indian Removal* (1932). Foreman discusses the Cherokee and the Trail of Tears in book 4.

5. On generally peaceful relations before the emigration era, see Robert M. Utley and Wilcomb E. Washburn, *Indian Wars* (first published in 1977, 2002 edition cited here) at page 157. The figure on Mandan deaths from smallpox is from the same source at page 146.

6. The 1851 Treaty of Fort Laramie can be viewed online at http://puffin.creighton.edu/lakota/1851_la.html. For a good overview of the 1851 and 1853 treaties, see Utley and Washburn, *Indian Wars*, at chapter VII, "The Whites Move In."

7. On Red Cloud's War, see generally James C. Olsen, *Red Cloud and the Sioux Problem* (1965). For general background on motivations for the 1867 Medicine Lodge Treaty and the Fort Laramie Treaty of 1868, see Donald Jackson, *Custer's Gold* (1966), at pages 7–8; Utley and Wilcomb, *Indian Wars*, at chapter VII, "The Whites Move In."

8. "[E]nded the Red Cloud War on Red Cloud's terms" is from Utley and Wilcomb at page 215. A copy of the 1868 Fort Laramie Treaty can be viewed online at http://puffin.creighton.edu/lakota/1868_la.html.

9. On public attention to the Custer expedition, see Jackson at page 23.

10. Descriptions of the departure from Fort Lincoln and the composition of the Custer expedition are drawn from Grinnell, *Memories*, at page 36; Grinnell, *Two Great Scouts* (1928), at pages 240–241; Jackson at pages 1, 17, and 25; and W.M. Wemett, "Custer's Expedition to the Black Hills in 1874," *North Dakota Historical Quarterly* (July 1932), pages 294–295.

11. Grinnell, *Memories*, page 35–36.

12. Ibid. at pages 35–37. Grinnell also discusses his admiration for Reynolds in his book *Two Scouts* (1928), at pages 17–18.

13. Wemett at page 296.

14. "An old miner and prospector" is from Grinnell, *Memories*, at page 36. For background on Custer's role in adding gold-hunting to his mission, see Jackson at 51. On Custer funding Ross and McKay, see Jackson at pages 81–82.

15. Jackson at page 3. There is debate about the authenticity of the inscribed stone, known as the "Thoen Stone" for the man who discovered it. Those wanting to judge for themselves can view the stone at the Adams Memorial Museum in Deadwood.

16. Background on pressure from the Dakota Territory and the passage from the *Press and Dakotaioan* are from Jackson at pages 4–9.

17. Grinnell, *Memories*, at pages 38–39.

18. "It seems that nothing ever died . . ." is quoted by Jackson at page 55.

19. On Grinnell's discoveries, see Jackson at pages 55–56. Custer's dispatch is quoted from a photocopy on file at the University of Montana's K. Ross Toole Archives.

20. Grinnell, *Memories*, at pages 38–39.

21. "General Custer was friendly . . ." and Grinnell's assessment of Custer's shooting and dogs are from *Two Great Scouts* at pages 241–242; "the impression of always . . ." is quoted from a Grinnell letter to W.H. Power, November 2, 1927, *Letter Book*, p. 626. For background on the *Letter Book*, see Chapter 1, note 1.

22. The entire campfire conversation concerning Black Hills gold is quoted from Grinnell's *From Notebooks of Black Hills Expedition* by John F. Reiger in *The Passing of the Great West: Selected Papers of George Bird Grinnell* (1972), at pages 99–101.

23. Curtis is quoted by Cleophas C. O'Harra, "The Discovery of Gold in the Black Hills," *The Black Hills Engineer* (November 1929), at pages 288–289.

24. Elizabeth Custer, *Boots and Saddles; or, Life in Dakota with General Custer* (originally published in 1885, 1961 edition cited), at page 200.

25. Custer's dispatch of August 2, 1874, is cited by numerous sources, including Jackson at pages 87–88.

26. Custer's dispatch of August 15, 1874, is cited from O'Harra at page 287.

27. *Press and Dakotaian* headlines cited by Jackson at page 89.

28. *Inter Ocean* story cited by Jackson at page 90 and O'Harra at page 289; "gold equivalent to $50 a day" is from *Harper's Weekly* (September 12, 1874) at page 753.

29. Cited by O'Harra at page 295.

30. Jackson at pages 116–118.

31. Theodore R. Davis, "The Buffalo Range," *Harper's New Monthly Magazine* (January 1869), at page 153.

32. On Sheridan and Sherman in the West, see Roy Morris, Jr., *Sheridan* (1992), at chapter 10, "General of the West."

33. On "The only good Indians . . ." see, e.g., Dee Brown, *Bury My Heart at Wounded Knee* (1970), at pages 170–172. For a somewhat tortured rationalization of the "only good Indians" affair, see Morris at pages 327–328. Debunking the story of Sheridan's testimony in Texas, see Dale Lott, *American Bison* (2002), at page 178.

34. Frank H. Mayer and Charles B. Roth (editors), *The Buffalo Harvest* (1958), at page 29.

35. "Kill every buffalo you can . . ." and "shoot it into a buffalo" are cited by Larry Barsness, *Heads, Hides & Horns* (1985), at pages 126–127.

36. Mayer quotes the unnamed officer at pages 29–30.

37. John Cook, *The Border and the Buffalo* (1907), at pages 132–133.

38. J. Wright Mooar, "Buffalo Days," *Holland's* (April 1933), at page 22. For the expression of similar sentiments by other buffalo hunters, see, e.g., John Cloud Jacobs, "The Last of the Buffalo," *The World's Work* (January 1909), at page 11,100, and Harlan B. Kauffman, "Hunting the Buffalo," *Overland Monthly* (August 1915), at page 170.

Chapter 7: "Ere Long Exterminated"

1. The quotation comes from the "Letter of Transmittal" dated June 1, 1876. The letter was attached as the cover to Grinnell's "Zoological Report," and both documents were part of William Ludlow's *Report of a Reconnaissance to the Yellowstone National Park* (1876). The author viewed a copy of the report at the Yellowstone National Park Archives in Gardiner, Montana.

2. As with "General" Custer, Ludlow's rank of colonel was a brevet designation from his Civil War service. His actual rank in 1875 was captain, but the title of *colonel* is used here, in part, for consistency with Grinnell's references to Ludlow.

3. George Bird Grinnell, *Memories*, at pages 41–42. For background on the *Memories* document, see chapter 1, note 3.

4. John F. Reiger, *The Passing of the Great West: Selected Papers of George Bird Grinnell* (1972), at page 109.

5. See Stallo Vinton, *John Colter: Discoverer of Yellowstone Park* (1926), at pages 38–42. Clark is cited from the *Journals of the Lewis and Clark Expedition* (volume 8), Gary E. Moulton (editor) at page 302.

6. On Colter's discovery of Yellowstone, see Vinton at chapter II; Aubrey L. Haines, *The Yellowstone Story* (1977), at volume 1, pages 35–38; and Hiram Martin Chittenden, *The Yellowstone National Park* (original edition, 1895; edition cited, 1917), at chapter IV, "John Colter."

7. Potts letter quoted by Haines at pages 41–42.

8. On Bridger and Yellowstone, including a lengthy analysis of "old Bridger's lies," see Haines at pages 52–59, 85, and Chittenden at chapter VI, "Bridger and His Stories."

9. On miners in Yellowstone, see Chittenden at chapter VIII, "The Gold-seeker;" and Haines at chapter 4, "Pick and Shovel Pilgrims." The "Indian catacomb" story is quoted by Haines at volume 1, page 83.

10. David E. Folsom, *The Folsom-Cook Exploration of the Upper Yellowstone in the Year 1869* (1894; preface by Nathaniel P. Langford), at page 9. Folsom's piece, viewed by the author in the Montana Collection at the University of Montana's Mansfield Library, was originally published by the Chicago-based *Western Monthly*, July 1870.

11. Folsom at pages 15–21.

12. Haines at pages 137–138.

13. On Cooke, the Northern Pacific Railroad, and Yellowstone, see ibid. at page 105.

14. Ibid. at pages 140–141. General Philip Sheridan also launched a party of army engineers to explore Yellowstone in the summer of 1871.

15. On Yellowstone's "creation myth," see in particular the insightful analysis of Paul Schullery and Lee H. Whittlesey, *Myth and History in the Creation of Yellowstone National Park* (2003).

16. "Public park or pleasuring ground . . ." is cited from *An Act to Set Apart a Certain Tract of Land Lying Near the Headwaters of the Yellowstone River as a Public Park*, Approved March 1, 1872.

17. On the politics of the creation of Yellowstone National Park, see H. Duane Hampton, *How the U.S. Cavalry Saved Our National Parks* (1971), at pages 27–31, and Haines at pages 165–173. On Moran, see generally Joni Louise Kinsey, *Thomas Moran and the Surveying of the American West* (1992). On Jackson, see generally Beaumont Newhall and Diana E. Edkins, *William H. Jackson* (1974).

18. Hampton at pages 30–31.

19. Grinnell, *Memories*, at page 42.

20. "[N]ot much in excess of one hundred miles" is from Ludlow at page 11.

21. Grinnell, *Memories*, at pages 45–46.

22. Ibid. at page 48.

23. "[S]aw and identified . . ." is from Grinnell's "Letter of Transmittal," an attachment to Colonel Ludlow's *Report of a Reconnaissance to the Yellowstone National Park*. Grinnell's discussion of the buffalo seen on the trip is from his "Zoological Report," another attachment to Ludlow's report. Grinnell's discussion of buffalo trails two feet deep is from Grinnell's letter to E. W. Nelson, May 11, 1916, *Letter Book*, page 116. For background on the *Letter Book*, see Chapter 1, note 1.

24. Ludlow at page 12.

25. In his memoirs, Grinnell refers his readers to Colonel Ludlow's official report for the day-by-day accounting of the expedition's time in the part. For the journey between Fort Ellis and Mammoth, including the description of the party's fishing, see Ludlow at pages 17–18.

26. George Bird Grinnell, "Letter of Transmittal," attached to Ludlow's *Report of a Reconnaissance to the Yellowstone National Park*.

27. Ibid. On the status of Grinnell's letter as one of the first official reports of its kind, see Reiger at page 119.

28. Ludlow, *Report of a Reconnaissance of the Black Hills,* at page 18. "The only blemishes ..." is from Ludlow, *Report of a Reconnaissance to the Yellowstone National Park,* at page 26.

29. Ludlow, *Report of a Reconnaissance to the Yellowstone National Park,* at pages 26–27.

30. The best historical overview of the development of environmental thought in the United States is Roderick Nash's classic, *Wilderness and the American Mind* (1967), on which much of this section is based.

31. James Fenimore Cooper, *The Prairie* (1827), at chapter VII. On the earlier development of Romantic writing in England and its impact on American thinking, see James C. McKusick, *Green Writing: Romanticism and Ecology* (2000).

32. Cooper, *The Prairie,* at chapter XXIII.

33. On the importance of Romantic artists in the development of appreciation for the beauty of wilderness, see Nash at pages 78–83.

34. On Moran, see generally Kinsey.

35. Nash at page 67.

36. On wilderness and American nationalism, see Nash at pages 67–69.

37. "Wonderful freaks of physical geography" is from Nathanial P. Langford, *Report of the Superintendent of the Yellowstone National Park for the Year 1872,* at page 7.

38. On the early failure to provide appropriations for Yellowstone National Park, see Hampton at chapter III, "The Early Years in Yellowstone." According to Hampton, the promise *not* to seek any appropriation for the park for at least one year had been a part of the bargaining to secure the necessary votes to bring the park into existence. See Hampton at page 34.

39. See Ludlow, "Letter of Transmission." Ludlow's recommendation for Department of War jurisdiction is at page 36.

Chapter 8: "A Weekly Journal"

1. J. Wright Mooar, "Buffalo Days," *Holland's* (May 1933), pages 11–12.

2. On Charlie Reynolds's death and Grinnell's near miss with Custer, see George Bird Grinnell, *Memories,* at pages 49–50. For background on the *Memories* document, see chapter 1, note 3.

3. Grinnell, *Memories,* at page 34.

4. The most comprehensive work examining the role of sportsmen as protectors of the environment has been done by John F. Reiger, professor of history at Ohio University–Chillicothe and the former executive director of the Connecticut Audubon Society. Reiger's research and his important book, *American Sportsmen and the Origins of Conservation* (1st edition, 1974; revised edition, 2001) is a major source for this section. Indeed Reiger deserves much of the credit for preserving the extensive personal papers of George Bird Grinnell. Grinnell bequeathed his papers to a friend named John P. Holman, who in turn passed them to the Connecticut Audubon Society's Birdcraft Museum. At one point, Reiger physically saved Grinnell's paper from being mistakenly thrown out as garbage. Reiger edited and

provided commentary to the Grinnell papers in a book, *The Passing of the Great West: Selected Papers of George Bird Grinnell* (1972), after using them as the basis for his doctoral thesis. In 1984, the papers were transferred to Yale University. The University of Montana has a copy of the Grinnell papers, used by the author in researching this book.

5. Charles Eliot Goodspeed, *Angling in America: Its Early History and Literature* (1939), at pages 73–76. See also Reiger, *American Sportsmen*, at pages 8 and 16.

6. Goodspeed at pages 132-133.

7. Jerome V.C. Smith, *Natural History of the Fishes of Massachusetts, Embracing a Practical Essay on Angling* (originally published in 1833, reprinted in 1970), at pages 345–346.

8. Robert Barnwell Roosevelt, *Superior Fishing; or, the Striped Bass, Trout, and Black Bass of the Northern States* (1865), at pages 19–21, 184.

9. For background on Marsh and samples of his writings, see David Lowenthal, *George Perkins Marsh: Prophet of Conservation* (1958; revised edition, 2000), and Stephen C. Trombulak (editor), *So Great a Vision: The Conservation Writings of George Perkins Marsh* (2001).

10. Trombulak at page 28.

11. Marsh's report is cited from Trombulak's collection of his writings, at pages 62–72.

12. On the broader significance of Marsh's fish report, see Lowenthal at pages 182–186. For a detailed discussion of Marsh, sportsmen, and the early development of environmental thought, see Reiger, *American Sportsmen*, at pages 22–29.

13. "[F]ountain-head of the conservation movement" is from Lewis Mumford, *The Brown Decades: A Study of the Arts in America, 1865–1895* (1955), at page 78. Stegner is quoted from his article, "It All Began with Conservation," *Smithsonian* (April 1990), at page 38.

14. George Perkins Marsh, *Man and Nature* (1864, reprinted 2003, David Lowenthal, editor), at page 11.

15. Marsh, *Man and Nature*, at chapter 1, footnote 35.

16. Marsh, *Man and Nature*, at page 35.

17. David Lowenthal, introduction to 2003 edition of Marsh's *Man and Nature*, at page xvi.

18. Reiger, *American Sportsmen*, at pages 56–57.

19. Donald W. Klinko, *Antebellum American Sporting Magazines and the Development of a Sportsmen's Ethic* (December 1986, Ph.D. dissertation, Washington State University), at pages iv–v.

20. Cited by Klinko at page 79.

21. Quotations from ibid. at pages 101 and 103.

22. See Klinko at pages 110, 199, 205, and 206.

23. Marsh, *Man and Nature*, at page 244.

24. See Klinko at 102, 111, and 113, and Reiger, *American Sportsmen*, at page 28.

25. On the impact of the Civil War, see Frank Luther Mott, *A History of American Magazines: 1850–1865* (1938), at pages 5–8. On the demise of *Spirit of the Times*, see ibid. at page 203.

26. Frank Luther Mott, *A History of American Magazines: 1865–1885* (1938), at page 5.

27. Cited by ibid. at page 210.

28. Reiger at page 3.

29. *Forest and Stream* (August 14, 1873), at page 8.

30. *Forest and Stream* (August 21, 1873), at page 24.

31. "Destruction of the buffalo," including discussion of the Colorado law is from *Forest and Stream* (October 16, 1873), at page 152. Calls for a season in which buffalo hunting is closed and for the cooperation of western governors are from *Forest and Stream* (January 22, 1874), at page 376.

32. *Forest and Stream* (December 25, 1873), at pages 305–306.

33. *Forest and Stream* (January 22, 1874), at page 376.

34. *Forest and Stream* (April 30, 1874), at page 184.

35. Grinnell, *Memories*, at page 51.

36. Ibid. at pages 53–54.

37. Grinnell, "A Trip to North Park," *Forest and Stream* (September 25, 1879), at pages 670–671.

38. Grinnell, "A Trip to North Park," *Forest and Stream* (October 2, 1879) at page 691.

39. Grinnell, "A Trip to North Park," *Forest and Stream* (October 30, 1879), at page 771.

40. Description of Hallock's alleged drinking problem and the corporate machinations at Forest and Stream Publishing Company are from Grinnell, *Memories*, at page 56.

Chapter 9: "No Longer a Place for Them"

1. Railroad remarks quoted by Edward W. Nolan, "Not Without Labor and Expense," *Montana* (summer 1983), at page 8.

2. George Bird Grinnell, "Introduction," *The Works of Theodore Roosevelt* (national edition, 1926), at pages xvi–xvii.

3. Grinnell, *Forest and Stream*, February 19, 1880, at page 51. Note: As with newspapers today, editorials in *Forest and Stream* were unsigned. As editor, though, Grinnell took personal responsibility for writing the editorial page. The use of pen names was also common; when attributed in *Forest and Stream*, for example, in his writings as a correspondent in the West, Grinnell often wrote as "Yo."

4. William Temple Hornaday, "Discovery, Life History, and Extermination of the Bison," *Report of the National Museum* (1887), page 510.

5. Victor Grant Smith (Jeanette Prodgers, editor), *The Champion Buffalo Hunter* (1997). For background on Smith's autobiography, see Prologue, note 1.

6. On Grinnell's move to *Forest and Stream*, see *Memories*, at page 56. Background on the *Memories* document can be found at chapter 1, note 3.

7. "Trout hogs" is from *Forest and Stream*, June 16, 1881; "Maine moose murders" is from *Forest and Stream*, October 13, 1881, at page 203; "law-defying marketmen and the challenge to the New York State Association for the Protection of Fish and Game is from *Forest and Stream*, June 17, 1880, at page 394.

8. *Forest and Stream* (November 8, 1883), at page 281.

9. *Forest and Stream* (February 12, 1880), at page 31.

10. On the Red Cloud incident, see Grinnell, "Sketch of Professor O.C. Marsh," *The Popular Science Monthly* (September 1878), at page 615; and William H. Goetzmann, *Exploration and Empire* (1966), at page 428–429.

11. Grinnell, *Forest and Stream* (February 12, 1880), at page 31.

12. Grinnell, *Forest and Stream* (May 11, 1882), at page 283.

13. Grinnell, "The Mountain Sheep and Its Range," *American Big Game and Its Haunts* (1904), at page 348.

14. See, e.g., "The Work of a State Game Protective Association," *Forest and Stream* (June 17, 1880), at page 394; "Big Game in Wyoming," *Forest and Stream* (April 27, 1882); "A Question of Ethics," *Forest and Stream* (February 16, 1882), at page 44; "Game Protection Fund," *Forest and Stream* (May 15, 1884), at page 301.

15. Grinnell, "American Game Protection, A Sketch," in *Hunting and Conservation: The Book of the Boone & Crockett Club* (1925), Grinnell and Charles Sheldon (editors), at pages 223–224.

16. Grinnell, *Forest and Stream* (June 29, 1882), at page 423.

17. "C.B.," *Forest and Stream* (January 27, 1881), at page 511.

18. *Forest and Stream* (March 25, 1880), at page 151.

19. T. S. Palmer, "Chronology and Index of the More Important Events in American Game Protection," U.S. Department of Agriculture, *Biological Survey Bulletin* Number 41 (GPO 1912), at pages 17–18. See also Barsness at pages 128–129.

20. A description of the bill, H.R. No. 921, and the bill's complete floor debate can be found in the *Congressional Record* (March 10, 1874), at pages 2105–2109. Congressman Fort originally introduced H.R. No. 921 on January 5, 1874 (*Congressional Record*, January 5, 1874, at page 371). Fort also introduced a second bill, H.R. No. 1689, that would have taxed buffalo hides (*Congressional Record*, February 2, 1874, at page 1123). For an unofficial legislative history of the 1874 effort to protect the buffalo, see Hornaday at pages 513–517.

21. *Congressional Record* (March 10, 1874), at page 2106.

22. Ibid. at page 2107.

23. Ibid.

24. Ibid. at page 2108.

25. Ibid. at page 2109.

26. See Hornaday at page 517.

27. Delano report cited by E. Douglas Branch, *The Hunting of the Buffalo* (1929), at page 176.

28. On the 1820 caravan, see Hornaday at page 487. On the impact of the Hudson's Bay Company on the settlement of western Canada, see Frank C. Roe, "The Extermination of the Buffalo in Western Canada," *The Canadian Historical Review* (March 1934), at pages 10–11. On the destruction of the herd in 1870s, see Roe at pages 12–13 and Hornaday at pages 504–505.

29. The term *Great Mother* is quoted by Robert M. Utley, *Indian Wars* (1977), at pages 265. On the U.S. policy of blocking the migration of buffalo into Canada, see, e.g., Roe at pages 16–17. "[C]ordon of half-breeds . . ." is from C.M. MacInnes, *In the Shadow of the Rockies* (1930), at pages 145–146. On the Canadian protests, see Joseph Howard Kinsey, *Strange Empire* (1952), at pages 292–293. On the "coup de grâce," see Branch at chapter XI; see also Larry Barsness, *Heads, Hides & Horns* (1985), at pages 127–128.

30. On the demise of the Canadian herd, see Roe at pages 16–20.

31. On the surrender of Sitting Bull, see Utley at pages 266–267.

32. Barsness at page 129.

33. *Sioux City Journal* quoted from Hornaday at page 503.

34. On the beginning of summer, hide hunting in the north, see Hornaday at page 506.

35. Lieutenant Partello is cited by Hornaday at page 509.

36. Ibid.

37. Mari Sandoz, *The Buffalo Hunters* (1954), at page 355.

38. Ibid. at pages 356–357. Vic Smith is quoted from Barsness at page 131.

39. Information on the Northern Pacific shipments is from Hornaday at page 513. Figures on boxcars are from information in Barsness at page 129.

40. Warner, Beers & Co., *History of Montana* (1885), at pages 30–31.

41. Barsness at pages 131–132. Cowboy Ed Carlson's story is from J.L. Hill, *Passing of the Indian* (1917), at pages 36–37.

42. On Roosevelt's motivation for going west, see James B. Trefethen, *Crusade for Wildlife* (1961), at page 2. Smith gives his account of hunting with Roosevelt in his book, *The Champion Buffalo Hunter*, at pages 128–131. The painting is discussed in a preface by the book's editor, Jeanette Prodgers, at pages xxiii.

43. Hornaday at pages 522, 529, and 541.

44. Ibid. at pages 540, 545, and 547.

45. Ibid. at page 537.

46. Ibid. at pages 542–543.

47. After removal from the Smithsonian in 1957, the Hornaday Bison Group was stored for decades at the University of Montana. In 1996, the restored diorama was placed on display at the Museum of the Northern Great Plains in Fort Benton, Montana, where it can be seen today. Information on the Hornaday Bison Group, including the Hornaday quotation, is from a paper presented at the museum's rededication ceremony: Douglas Coffman and Eugene Oregon, *Art to Eternity: A Story of Hornaday's Bison Group* (June 30, 1996).

48. Grinnell, *Sportsman's Gazetteer and General Guide* (1877), pages 34–35. The nominal author of this book, a publication of the Forest and Stream Publishing Company, was the magazine's first editor, Charles Hallock. However, as Grinnell makes clear in his memoirs, it was he (uncredited) who wrote the sections of the book covering mammals and birds. See Grinnell, *Memories*, at page 50. See also John F. Reiger, *American Sportsmen and the Origins of Conservation* (2001 edition), at pages 132–133.

49. On former buffalo hunters, see Barsness at page 131.

50. On the market for buffalo trophy mounts, see letter from Captain George S. Anderson to the Secretary of Interior (May 24, 1894), on file at the Yellowstone National Park Archives, Gardiner, Montana, Letter Box 1, document 286.

Chapter 10: "Blundering, Plundering"

1. John Muir, *Our National Parks* (1901), at page 30.

2. General historical information on Yellowstone National Park in this chapter is based largely on information from three sources: The most comprehensive history of the park is Aubrey L. Haines' two-volume treatment, *The Yellowstone Story: A History of Our First National Park* (1977). The first comprehensive history of the park, still highly readable today, is Hiram Martin Chittenden's *The Yellowstone National Park* (first edition, 1895; author cites to second edition, 1917 printing). One of the most interesting treatments of the politics machinations involving the park—with a particular emphasis on conservation—is H. Duane Hampton's *How the U.S. Cavalry Saved Our National Parks* (1971). On Jay Cooke and Cooke City, see Haines, volume I, at page 267.

3. Praising the removal of squatters, see *Forest and Stream*, January 8, 1885, at page 461. "The new Superintendent . . ." is from *Forest and Stream*, March 12, 1885, at page 121.

4. "[E]ither the height of audacity . . ." is from *Forest and Stream* (May 14, 1885), at page 305. On Carpenter's scheme, see also *Forest and Stream* editorials of April 9, 1885, at page 201 and April 23, 1885, at page 245.

5. The entire "no wind in Livingston" episode is related gleefully in *The Livingston Enterprise* (February 2, 1885), at page 3. The *Enterprise* did offer Carpenter an opportunity to rebut the charges against him (unconvincingly) on April 18, 1885, at page 3.

6. "[A] stampede . . ." is from ibid.; see also Haines at pages 315–316.

7. "[E]nrolled himself among the number . . ." is from *Forest and Stream* (April 23, 1885), at page 245.

8. Haines at pages 144 and 196; "impromptu village" is from the *Rocky Mountain Weekly Gazette*, July 24, 1871, cited by Haines in footnote 129 of chapter 5. On the "national swimming school," see Hampton at page 36.

9. *An Act to Set Apart a Certain Tract of Land Lying Near the Headwaters of the Yellowstone River As a Public Park,* approved March 1, 1872. Codified at 16. U.S.C. sections 21–22.

10. Haines at pages 180–181.

11. Emphasis added. The original rules are printed in the *Report of the Superintendent of the Yellowstone National Park for the Year 1877*, at page 993 (on file at the Yellowstone National Park Archives in Gardiner, Montana).

12. *Report of the Superintendent of the Yellowstone National Park for the Year 1872*, at pages 4–5 (on file at the Yellowstone National Park Archives in Gardiner, Montana.)

13. John Gray, "The Last Rights of Lonesome Charlie Reynolds," *Montana* (summer 1963), at page 43.

14. Norris's *Report of the Superintendent of the Yellowstone National Park for the Year 1877*, at

page 837, and *Report of the Superintendent of the Yellowstone National Park for the Year 1878*, at page 993 (both on file at the Yellowstone National Park Archives in Gardiner, Montana).

15. *Report of the Superintendent of the Yellowstone National Park for the Year 1877*, at pages 842–843.

16. *Report of the Superintendent of the Yellowstone National Park for the Year 1878*, at page 994. See also Hampton at pages 45–47.

17. "Report of the Gamekeeper," included as part of *Report of the Superintendent of the Yellowstone National Park for the Year 1880*, at page 50 (on file at the Yellowstone National Park Archives in Gardiner, Montana).

18. *Forest and Stream* (January 4, 1883), at page 441.

19. *Forest and Stream* (January 11, 1883), at page 461.

20. "The People's Park" is found at *Forest and Stream* (January 18, 1883), at page 481. On Grinnell and the development of a national park concept, see also John F. Roger, *American Sportsmen and the Origins of Conservation* (1975, 2001), at chapter 5, "Development of the National Park Concept."

21. "It is the duty . . ." is from *Forest and Stream* (January 18, 1883), at page 481. "The Park is for rich and poor . . ." is from *Forest and Stream* (January 11, 1883), at page 461.

22. For a reprint of Secretary Teller's letter and for Grinnell's "The Park Saved," see *Forest and Stream* (January 25, 1883) at page 501.

23. Ibid. For further background on the Yellowstone Park Improvement Company, see Haines at chapter 11.

24. "[T]he main trouble . . . " is from *Forest and Stream* (August 5, 1880), at page 3; "again stultified . . . " is from *Forest and Stream* (December 14, 1882), at page 382. "Rules and regulations . . . " is from *Forest and Stream*, March 1, 1883, at page 81.

25. Hiram Chittenden, *The Yellowstone National Park* (first edition, 1895; 1917 edition cited), at page 109.

26. "[T]est the power of the Geysers . . . " is quoted by Hampton at page 56. On Sheridan's forest fire, see Haines at page 263. On firewood, see Haines at page 306.

27. The "Buffalo Hunting" stereoscope card is on file at the Yellowstone National Park Archives in Gardiner, Montana. On tourist hunting and the lack of game along trails, see Haines at pages 200, 207–208.

28. Haines at pages 303–304.

29. *Forest and Stream* (March 13, 1883), at page 121. See also Hampton at page 61.

30. "Rabbit catchers" is from Haines at chapter 11; the other critics are quoted by Hampton at page 66. See also *The Livingston Enterprise* (April 18, 1885), at page 2.

31. "[L]aughed at" is from Jas. H. Dean, Letter to P.H. Conger (July 31, 1883), *Letters Received Book,* file number 1367, Yellowstone National Park Archives.

32. "I would not ask" is quoted by Haines at pages 302–303; see also *Report of the Superintendent of the Yellowstone National Park* (1883), at pages 4–5 (on file at the Yellowstone National Park Archives in Gardiner, Montana).

33. "[S]ome lawless men know . . . " is from quoted by Haines at page 307.

34. Letter from Edmund Fish to P. H. Conger (June 4, 1885), *Letters Received Book,* file number 1420, Yellowstone National Park Archives. "Let the Military . . . " is cited by Hampton at page 67.

35. William Ludlow, *Report of a Reconnaissance to the Yellowstone National Park* (1876), at page 36. The author viewed a copy of the report at the Yellowstone National Park Archives.

36. Sheridan is quoted by Hampton at page 55.

37. *Forest and Stream* (December 14, 1882), at page 383.

38. On the Vest bill and its House companion, see Hampton at page 56 and Haines at pages 268–269.

39. Language cited in *Forest and Steam* (March 8, 1883), at page 101.

40. "[S]trange apathy . . . " and "It is very evident . . . " are from *Forest and Stream* (February 22, 1883), at page 61; ibid.; "some of the sharpest intellects . . . " and "unlimited money . . . " are from *Forest and Stream* (March 8, 1883), at page 61.

41. *Forest and Stream* (December 14, 1882), at page 383.

Chapter 11: "The Meanest Work I Ever Did"

1. Cited by H. Duane Hampton, *How the U.S. Cavalry Saved Our National Parks* (1971), at page 61.

2. *Forest and Stream* (December 14, 1882), at page 382.

3. George Bird Grinnell, *Brief History of the Boone and Crocket Club* (1910), at page 16.

4. Much of the material in this section, including "Third House of Congress" and the figures for Texas and Pacific Railway lobbyists, is based on information from David J. Rothman's *Politics and Power: The United States Senate, 1869–1901* (1966), at chapter VII, "The Business of Influence."

5. Rothman at page 208.

6. Ibid. at page 209.

7. Ibid. at page 196, citing correspondence between Central Pacific officials.

8. Ibid. at 213.

9. Ibid. at pages 200, 217–218.

10. Ibid. at page 208.

11. Letter from Grinnell to Luther H. North (February 17, 1887), *Letter Book* at page 70. For background information on the *Letter Book* see chapter 1, note 1.

12. "[T]he interest of the people . . . " is from *Forest and Stream* (January 18, 1883), at page 481; on the people's ownership of federal resources, see "The People's Park," *Forest and Stream* (January 18, 1883), at page 481; "grandchildren of the present generation" is from *Forest and Stream* (December 14, 1882), at page 382.

13. Hampton at page 4.

14. "[R]easons why the forests should be preserved . . . " is from *Forest and Stream* (December 13, 1883), at page 381; on forestry, see also *Forest and Stream* (January 31, 1884) at page 2 and *Forest and Stream* (January 29, 1885) at page 2.

15. On Grinnell's pragmatism, see John F. Reiger, *American Sportsmen and the Origins of Conservation* (1975, 2001), at pages 113–118.

16. "The Mutual Interests of Farmers and Sportsmen" is from *Forest and Stream* (March 11, 1880), at page 110.

17. "A sentiment against . . . " is from *Forest and Stream* (October 18, 1883), at page 222. On the Wyoming Stock Growers' Association, see *Forest and Stream* (December 30, 1880) at page 423.

18. *Forest and Stream* (August 5, 1893), at page 93.

19. *Forest and Stream* (May 15, 1884), at page 301.

20. For Grinnell's *Forest and Stream* writings about birds, see, e.g., March 23, 1882 (at page 143); March 30, 1882 (at page 163); and August 7, 1884 (at page 21).

21. *Forest and Stream* (February 11, 1886), at page 41. Membership numbers for 1887 are from *Audubon Magazine* (February 1887) at page 5. The early organization and the subsequent history of the Audubon Society are somewhat complex. By 1889, the organization had become so successful that Grinnell could no longer manage it while still assuming the responsibilities of *Forest and Stream*. He discontinued his national role. Local and state societies continued, however, and in 1905 a new National Association of Audubon Societies was formed. Grinnell would serve as director of this reconstituted body. On the history and formation of the Audubon Society, see Reiger at pages 98–101.

22. Grinnell, "Introduction to Volume 1," *The Works of Theodore Roosevelt* (national edition, 1926), at page xiii.

23. *Forest and Stream* (July 2, 1885), at page 451.

24. Grinnell, *The Works of Theodore Roosevelt*, at pages xiv–xv. See also Reiger at pages 146–150.

25. Grinnell, *The Works of Theodore Roosevelt*, at page xv.

26. "[W]as always fond . . . " is from ibid. at pages xv–xvi. On Roosevelt and Mayne Reid, see Reiger at page 148.

27. Grinnell, *The Works of Theodore Roosevelt*, at page xvi.

28. According to Grinnell, Roosevelt first had the idea for "establishment of a club of amateur riflemen, an organization which should be made up of good sportsmen who were also good big-game hunters." Grinnell, *The Works of Theodore Roosevelt*, at page xvii.

29. Grinnell, *Brief History of the Boone and Crockett Club*, at page 3. *Minutes of Boone and Crockett Club Meeting of February 29, 1888*, viewed by the author at the Boone and Crockett National Headquarters in Missoula, Montana. On Boone and Crockett's status as the first national group formed to promote conservation legislation, see Reiger at page 153.

30. Boone and Crockett's founding principles can be found in Grinnell's *Brief History of the Boone and Crockett Club*, at page 4.

31. *Minutes of Boone and Crockett Club Meeting of February 29, 1888.*

32. "It would seem . . . " is from *Forest and Stream* (February 16, 1888), at page 61; "names carried weight" is from Grinnell, *The Works of Theodore Roosevelt*, at page xviii.

33. Grinnell, *Brief History of the Boone and Crockett Club*, at page 20.

34. *Forest and Stream* (January 17, 1889), at page 513.

35. Aubrey L. Haines, *The Yellowstone Story* (1977), volume one, at pages 270–271.

36. Except where indicated, information and quotations on the Wyoming law are drawn from Haines at pages 311–312 and Hampton at page 68.

37. "[P]ut a club in my hands" is quoted by Haines at page 312–313; on the arrest by Ed Wilson, see Haines at page 321.

38. Ibid. at page 314. David Wear is quoted from the *Report of the Superintendent of the Yellowstone National Park* (1886), at page 4.

39. Hampton at pages 72–73.

40. The description of the Payson farce is based on Haines at pages 321–323 and Hampton at pages 72–73.

41. On the "Dogberry" commentary, Haines cites the *Livingston Enterprise* (August 15, 1885). Further coverage of the Payson affair can be found in the *Livingston Enterprise* editions of August 22, 1885, at page 2 and August 29, 1885, at page 3.

42. Haines at volume II, at page 4.

Chapter 12: "A Terror to Evil-Doers"

1. *Forest and Stream* (May 30, 1889), at page 373.

2. Background information on the transfer in Yellowstone from civilian to military administration, except where noted, is from H. Duane Hampton, *How the U.S. Cavalry Saved Our National Parks* (1971), at page 61, and Aubrey L. Haines, *The Yellowstone Story* (1977), volume one, at pages 323–326, and volume two, at pages 3–4.

3. *Forest and Stream* (February 20, 1980), at page 81.

4. For Senator Plumb's remarks, see *Congressional Record*, 49th Congress, 1st Session, XCII, Part 8, at page 7842. For Representative Reagan's remarks, see *Congressional Record*, 49th Congress, 1st Session, XVII, Part 8, at pages 7866–7867.

5. D.W. Wear, *Report of the Superintendent of the Yellowstone National Park* (August 20, 1886), at page 5. See also Captain Moses Harris, *Report of the Superintendent of the Yellowstone National Park* (October 4, 1886), at page 6, and Captain Harris, *Report of the Superintendent of the Yellowstone National Park* (June 1, 1889), at page 10.

6. A listing of the citations for the soldiers awarded the Congressional Medal of Honor can be found at http://www.army.mil/cmh-pg/mohciv.htm.

7. Telegram cited by Hampton at page 82.

8. Captain Moses Harris, *Report of the Superintendent of the Yellowstone National Park* (October 4, 1886), at page 7. See also Hampton at pages 82–83.

9. Order No. 5 is included in Captain Harris's *Report of the Superintendent of the Yellowstone National Park* (October 4, 1886), at pages 11–12. The Secretary of Interior's *Rules and Regulations of the Yellowstone National Park* (April 4, 1887) are on file at the Yellowstone National Park Archives in Gardiner, Montana.

10. Harris, *Report of the Superintendent of the Yellowstone National Park* (August 20, 1887), at page 3.

11. *Forest and Stream* (May 30, 1889), at page 373.

12. See, e.g., Harris, *Report of the Superintendent of the Yellowstone National Park* (June 1, 1889), at page 12. See also Captain George Anderson, "Protection of the Yellowstone National Park," *Hunting in Many Lands* (edited by George Bird Grinnell and Theodore Roosevelt, 1895), at pages 384–385.

13. Harris, *Report of the Superintendent of the Yellowstone National Park* (August 20, 1887), at page 3.

14. Ibid.

15. On skiing in the era, see Emerson Hough, "Winter in the Rockies," *Chicago Tribune*, December 23, 1894.

16. On Schwatka's ill-fated winter expedition, see Haines, volume 2, at pages 6–10.

17. Harris, *Report of the Superintendent of the Yellowstone National Park* (August 20, 1887), at page 4.

18. Ibid. at Appendix B.

19. Harris, *Report of the Superintendent of the Yellowstone National Park* (August 20, 1887), at page 13.

20. On the two winter "snowshoe" expeditions of 1888, see Harris, *Report of the Superintendent of the Yellowstone National Park* (August 15, 1888), at pages 5–6.

21. Ibid. at page 6.

22. Ibid. at page 3.

23. *Forest and Stream* (May 30, 1889), at page 373.

24. Harris, *Report of the Superintendent of the Yellowstone National Park* (June 1, 1889), at page 7.

25. On the sixteen expulsions, see Hampton at page 90. "The Penalties for violation . . . " is from Harris, *Report of the Superintendent of the Yellowstone National Park* (June 1, 1889), at page 10.

26. Harris's descriptions of his foes are from *Report of the Yellowstone National Park* (1887), at page 5 and *Report of the Yellowstone National Park* (1888), at page 13. "All sorts of worthless . . . " is from *Report of the Yellowstone National Park* (October 4, 1886), at page 11.

27. On the Garfield incident, see Haines at volume 2, page 26; Harris, *Report of the Yellowstone National Park* (1889), at page 4.

28. The figure of eighty buffalo is from Anderson, "Protection of the Yellowstone National Park," at pages 414–415.

29. The description of the stage-robbing incident is based on Haines, volume 2, at pages 12–14.

30. Harris, *Report of the Yellowstone National Park* (1888), at page 4. *Forest and Stream* (January 30, 1890), at page 21.

Chapter 13: "A Single Rock"

1. *Forest and Stream* (December 14, 1882), at page 383.

2. Information on the buffalo bone industry is drawn primarily from LeRoy Barnett,

"Ghastly Harvest: Montana's Trade in Buffalo Bones," *Montana* (summer 1975), pages 2–13, and Wayne Gard, *The Great Buffalo Hunt* (1959), at chapter 18, "Bones on the Prairie." Gard also provides details of the 1870s bone industry on the plains of Kansas.

3. Hornaday cited by Barnett at page 4; "first cash crop" is cited by Barsness, *Head, Hides, & Horns*, at page 134.

4. The $7 per ton of bones is an 1884 figure from Barnett at page 4. Gard, at page 298, estimates that it took approximately 100 buffalo to generate a ton of bones. Smith's quotes are from Victor Grant Smith (Jeanette Prodgers, editor), *The Champion Buffalo Hunter* (1997), at page 132.

5. Long extracts of the Vest speech are cited in *Forest and Stream* (May 19, 1892), at page 474.

6. "I must say that I am sick . . . " is quoted from a letter from Grinnell to W.H. Phillips (December 15, 1892)—see *Letter Book*, at page 648; "destroy one of the great game ranges of the Park" is from *Forest and Stream* (April 3, 1890), at page 205; "a grave injury . . . " is from *Forest and Stream* (October 13, 1892), at page 309. For background information on the *Letter Book* see Chapter 1, note 1.

7. *Forest and Stream* (December 8, 1892), at pages 485–488. In the midst of a year that appeared to mark a fateful shift to Park opponents, Grinnell got a break from an unlikely source—the Northern Pacific Railroad Company. In the same article, Grinnell explained how, after a decade of operating behind the scenes in support of a railroad through the Park, the Northern Pacific had come to a startling conclusion: The ore in Cooke City was not worth the expense—too low a grade to justify a railroad. Testifying before a House committee, Northern Pacific President Thomas Oakes stated flatly, "There is nothing in the Cooke City mines, and we did not want a railroad there." The Northern Pacific's former proxy, the Montana Mineral Railroad Company, would continue to push for the link, as would powerful real-estate speculators with a vested interest in a railroad through the Park. But it was Northern Pacific that had served as the most powerful force behind the segregation effort, and its change of position was a blow to advocates of segregation.

8. "[C]ommand center" is quoted from Grinnell, *Memories*, at page 62. For background on the *Memories* document, see chapter 1, note 3.

9. *Forest and Stream* (December 15, 1892), at page 507.

10. Roosevelt's letter was published in *Forest and Stream* (December 15, 1892), at page 514.

11. Grinnell reported details of the meeting in *Forest and Stream* (January 22, 1891), at page 3.

12. Hague, like Vest, would disappoint Grinnell by supporting segregation as a necessary evil to win enforcement legislation for the park.

13. *Forest and Stream* (December 8, 1892), at page 486.

14. *Congressional Record,* May 10, 1892, at page 4125.

15. *Congressional Record,* December 14, 1886, at page 153.

16. *Congressional Record,* December 14, 1886, at page 150.

17. *Forest and Stream* (February 23, 1893), at page 155.

18. *Forest and Stream* (May 9, 1889), page 313. *Forest and Stream* (January 22, 1891), at page 1.

19. General biographical information on Anderson as well as the Grinnell quote are from Haines, volume 2, at pages 454–455.

20. See Hampton at pages 111–112.

21. Captain George Anderson, "Protection of the Yellowstone National Park," *Hunting in Many Lands* (1895), at page 401.

22. Hampton at page 108.

23. Ibid. at pages 389–390.

24. McDermott letter (September 14, year not given) and Krachy letter (possibly May 1893) are part of the collection at the Yellowstone National Park Archives in Gardiner, Montana.

25. The Robins letter (December 24, 1893) and the Marshall letter (August 22, 1895) are part of the collection at the Yellowstone National Park Archives in Gardiner, Montana.

26. Correspondence between Anderson and Grinnell as well as the letter from Secretary Noble to Anderson (June 13, 1893) are part of the collection at the Yellowstone National Park Archives in Gardiner, Montana.

27. F. Lusk letter (January 25, 1892) and Veritas letter (September 5, 1895) are part of the collection at the Yellowstone National Park Archives in Gardiner, Montana.

28. Anderson, "Protection of the Yellowstone Park," at page 395. See also Haines, volume 2, pages 60–62.

29. The account of Van Dyke's capture is from ibid. at pages 385–386.

30. On the bonfire, see Anderson, "Protection of the Yellowstone National Park," at page 395. On the horse, see *Letter from Frank Chatfield to Secretary of Interior John W. Noble* (November 24, 1892), viewed by the author at the Yellowstone National Park Archives in Gardiner, Montana. "Every bit of property ... " and "confiscation if made" are from Anderson, *Report of the Superintendent of the Yellowstone National Park* (August 15, 1892), at page 9.

31. See, e.g., Letter from Anderson to the Secretary of Interior (September 14, 1893), Yellowstone National Park Archives, *Letters Sent* book at pages 333–334.

32. On Van Dyke, see Letter from Captain F.A. Boutelle to Van Dyke (December 2, 1890), Yellowstone National Park Archives, *Letters Sent* book at pages 199–200. "[S]imple removal" is quoted by Hampton at pages 109–110 from a telegram sent by Anderson to the Secretary of Interior (August 10, 1892).

33. "[F]or some reason ... " is from Anderson, "Protection of the Yellowstone National Park," *Hunting in Many Lands*, at pages 395–396.

34. The secretary's order to release the prisoner and "This imprisonment of a month ... " are cited by Hampton, at pages 109–110.

35. On $500 for mounted buffalo heads and on London prices, see Anderson Letter to the Secretary of Interior (May 24, 1894), Yellowstone National Park Archives, *Letters Sent* book at page 58. The poacher advertising is reported in the Veritas Letter at pages 3–4.

36. E.S. Thompson's letters to the Secretary of the Interior (September 25, 1894) and to Captain Anderson (October 23, 1894) are part of the collection at the Yellowstone National Park Archives.

37. Barnett at pages 5, 13.

Chapter 14: "For All It Is Worth"

1. Letter on file at the Yellowstone National Park Archives, Gardiner, Montana.

2. Descriptions of Howell are from *Forest and Stream* correspondent Emerson Hough, *Forest and Stream* (May 5, 1894), at page 378.

3. Captain Anderson didn't believe Howell's story about fighting with Noble. In Anderson's opinion, Howell sent Noble south with their first load of heads. Ibid. The term *hot ground* is from Emerson Hough, "Winter in the Rockies," *Chicago Tribune* (December 24, 1894).

4. Hough, *Forest and Stream* (May 5, 1894), at page 378.

5. "[B]rave as Caesar . . . " is quoted from "Protection of the Yellowstone Park," *Hunting in Many Lands* (1895), at page 385. See also Anderson, *Report of Superintendent of Yellowstone National Park* (August 15, 1891), at page 9; Aubrey Haines, *The Yellowstone Story* (1977), volume 2, at page 448.

6. The toe story is quoted from E. Hough, *Forest and Stream* (May 5, 1894), at page 378. On Burgess's service in the Indian wars, see Haines at page 445.

7. "[S]igns of bison" and "old and could not be followed" are from Anderson, *Hunting in Many Lands*, pages 396–397. On Burgess's broken ax, see *Forest and Stream* (May 5, 1894), at page 377.

8. *Congressional Record*, at September 6, 1893, page 1271; September 9, 1893, at page 1341. See also Hampton at page 118.

9. George Bird Grinnell, Letter to Captain George Anderson (December 13, 1893), part of the collection at the Yellowstone National Park Archives.

10. H. Duane Hampton, *How the U.S. Cavalry Saved Our National Parks* (1971), at pages 118–119.

11. Grinnell, Letter to Madison Grant (January 4, 1918), viewed as part of *Letter Book*. For background on the *Letter Book* see Chapter 1, note 1.

12. *Forest and Stream* (March 24, 1894), page 241. See also Hampton at pages 118–121.

13. *Brief History of the Boone and Crockett Club* (1910), at page 14. Viewed by the author at Boone and Crockett Club headquarters, Missoula, Montana.

14. The account of the pursuit and arrest of Howell is based primarily on information from the following sources: Anderson, "Protection of the Yellowstone Park," *Hunting in Many Lands* (Roosevelt and Grinnell, editors, 1895), at pages 396–400; E. Hough, "*Forest and Stream*'s Yellowstone Park Game Exploration," *Forest and Stream* (May 5, 1894), at pages 377–378; and E. Hough, "Winter in the Rockies," *Chicago Tribune* (December 23, 1894).

15. "Calculating the time . . . " is from Anderson, "Protection of the Yellowstone Park" at

page 397; "with a whole ax . . . " is from E. Hough, *Forest and Stream* (May 5, 1894), at page 377; "desperate character" and "resist arrest . . . " are from Anderson to the Secretary of Interior (March 17, 1894), Yellowstone National Park Archives, *Letters Sent* book at pages 1–9.

16. Anderson requested reimbursement from the Department of Interior for this amount in a letter of October 29, 1894 (on file at the Yellowstone National Park Archives). He was turned down.

17. In some accounts, Troike is identified as a sergeant.

18. "[A] severe storm" is quoted from Anderson, "Protection of the Yellowstone Park," at page 397.

19. Quotes from scout Felix Burgess are from an interview conducted by E. Hough and published in *Forest and Stream* (May 5, 1894), at page 378.

20. Howell is quoted by Anderson, "Protection of the Yellowstone Park," at page 399.

21. E. Hough, quoting Burgess, *Forest and Stream* (May 5, 1894), at page 378.

22. E. Hough, *Forest and Stream* (May 5, 1894), at page 378.

23. Ibid. at page 377.

24. "[S]ock him just as far . . . " and "the Secretary of Interior . . . " are from E. Hough, quoting Anderson, *Forest and Stream* (May 5, 1894), at page 377.

25. *Forest and Stream* (April 14, 1894), at page 309.

26. Haynes, it will be remembered, recorded the first winter photographs of Yellowstone during a perilous exploration in 1887. On his presence with Hough, see Hampton at page 123.

27. Howell quoted by Hough, *Forest and Stream* (May 5, 1894), at page 378.

28. *Forest and Stream* (March 24, 1894), at page 241.

29. *Memories* at page 61. For background on the *Memories* document, see chapter 1, note 3.

30. For Haynes' photos, see *Forest and Stream* (May 5, 1894), at page 378.

31. "[W]rite his Senator and Representative . . . " is quoted from *Forest and Stream* (April 14, 1894), at page 309; "thousands upon thousands" is quoted from Grinnell, *Memories*, at page 62.

32. Lacey introduced the bill (H.R. 6442) on March 26, 1894, and it was referred to as the committee on Public Lands. See *Congressional Record*, 53rd Congress, 2nd Session, XXVI, March 26, 1894, at page 3252.

33. "[I]nfernal scoundrel" is from a Roosevelt letter to Anderson of April 30, 1894. "A man like that . . . " is from a Roosevelt letter to Captain George S. Anderson of March 30, 1894; "had made the greatest mistake . . . " is from a William Hallet Phillips letter to Captain George S. Anderson (March 31, 1894). Letters on file at the Yellowstone National Park Archives in Gardiner, Montana.

34. Theodore Roosevelt to Captain George S. Anderson (March 30, 1894). Letter on file at the Yellowstone National Park Archives in Gardiner, Montana.

35. Captain George S. Anderson, "Protection of the Yellowstone Park," *Hunting in Many Lands* (1895), at page 399.

36. "[B]e made the occasion . . . " is quoted from Anderson to the Secretary of Interior, March 17, 1894, Yellowstone National Park Archives, *Letters Sent* book at pages 1–9. Both Anderson and Smith letters are on file at the Yellowstone National Park Archives. See also Hampton at page 122.

37. *Congressional Record*, May 2, 1894, at page 4315.

38. *Forest and Stream* (April 14, 1894), at pages 309, 313. "At this rate . . . " is quoted from *Forest and Stream* (March 24, 1894), at page 241.

39. *Protection of Game in Yellowstone National Park*, House of Representatives Report No. 658, to Accompany H.R. 6442 (53rd Congress, 2nd Session, April 4, 1894).

40. The National Park Protective Act is cited from *Hunting in Many Lands* at pages 433–438.

41. *Hunting in Many Lands* at page 403.

42. Hampton at page 129.

Chapter 15: "Simple Majesty"

1. Coolidge speech quoted from the *New York Times* (May 16, 1925), at page 23.

2. "Buffalo in the United States are doomed" is quoted from a Grinnell letter to Anderson dated December 28, 1896. This letter as well as Anderson's ongoing correspondence with his informants are part of the collection at the Yellowstone National Park Archives in Gardiner, Montana.

3. The number of buffalo in the Park in 1902 is based on the number actually counted by park officials. Thus, it is possible that some animals escaped the count. Even accounting for these animals, the total appears unlikely to have been less than 50. On historic numbers, see *The Bison of Yellowstone*, National Park Service Monograph No. 1 (last updated January 24, 2005), at chapter 3. The monograph is available online at http://www.cr.nps. gov/history/online_books/bison/contents.htm.

4. Where not specifically noted, information about events in the following section is drawn primarily from four sources: John Kidder, "Montana Miracle: It Saved the Buffalo," *Montana* (spring 1965), at page 52; Paul Schullery, "'Buffalo' Jones and the Bison Herd in Yellowstone: Another Look," *Montana* (summer 1976), at page 40; Larry Barsness, *Heads, Hides, & Horns* (1985), at pages 155–157; Dale F. Lott, *American Bison* (2002), at pages 185–187; and Robert Easton and MacKenzie Brown, *Lord of the Beasts: The Saga of Buffalo Jones* (1961), at chapter 15.

5. "[B]alloon-propelling device" is quoted from Schullery at page 49; "patented water elevator" is quoted from Barsness at page 176.

6. Charles Jesse Jones, *Buffalo Jones' Forty Years of Adventure* (1889), at page 235.

7. The details concerning Walking Coyote's buffalo hunt are from Tom Jones, *The Last of the Buffalo* (1909), at pages 3–4.

8. Easton and Brown at page 262, footnote 5. See also *Annual Report of the Superintendent of Yellowstone Park* (October 15, 1902).

9. Barsness at page 176.

10. "[P]ompous, opinionated . . . " is quoted by Barsness at page 156.

11. The best summary of Jones's many enemies is the Schullery article. On Jones and Roosevelt in 1903, see Shullery at pages 44–46.

12. In this section, information on the numbers of Yellowstone buffalo and the techniques used to manage them are drawn primarily from *The Bison of Yellowstone* at chapter 3.

13. On the Canadian herd, see Frank G. Roe, "The Extermination of the Buffalo in Western Canada," *The Canadian Historical Review* (March 1934), at pages 16–20; Kidder at pages 57–66; Valerius Geist, *Buffalo Nation: History & Legend of the North American Bison* (1996), at pages 94–98.

14. Information on Hornaday and the American Bison Society is drawn from Lott at 187–188; Barsness at pages 162–164.

15. On the efforts to create buffalo reserves outside of Yellowstone, see David A. Dary, *The Buffalo Book* (1974, 1989), at chapter XVI, "Those Who Saved the Buffalo."

16. *Scribner's Magazine* (September 1892), at pages 267–268.

17. According to John F. Reiger, *American Sportsmen and the Origins of Conservation* (newest edition, 2001), at page 187, the quality of *Forest and Stream* declined rapidly after Grinnell left the helm. In 1915, the journal reduced its publication schedule from weekly to monthly, and in 1930 it ceased publishing altogether. The baton, though, did not fall. *Forest and Stream* sold its subscriber list to *Field & Stream*, which continues the great tradition of a thoughtful sportsmen's journal to this day.

18. Where not specifically noted, general biographical information in this section is based on Grinnell's obituaries, including the *New York Times* (April 12, 1938), at page 23; the *New York Herald Tribune* (April 12, 1938), at page 14; and the *Journal of Mammalogy* (August 18, 1938), at page 397.

19. Nor was Grinnell's interest in the Indians merely academic. He served as an advocate for many of the tribes he studied, including a successful campaign to remove a corrupt agent from the Blackfoot reservation. See, e.g., Gerald A. Diettert, "Grinnell's Glacier" (master's thesis, 1990), on file at the University of Montana, at page 76.

20. Coolidge speech quoted from the *New York Times* (May 16, 1925), at page 23.

21. See Schultz's account *Forest and Stream* (December 3, 1885), at pages 362–363; Diettert at pages 23–24.

22. *Forest and Stream* (December 17, 1885), at page 402.

23. Letter from Grinnell to C.C. Buell, *Century Magazine* (May 25, 1892), on microfilm at the K. Ross Toole Archives, Mansfield Library, University of Montana, cited by Diettert at page 87.

24. "I have been running around . . . " is cited by Diettery at page 94. "Paradise will be invaded . . . " is quoted from a Grinnell letter to George H. Gould (April 23, 1894); "make a bigger camp than Butte" is quoted by Grinnell in his diary (October 29, 1891). Copies of the above documents are available on microfilm at the K. Ross Toole Archives, Mansfield Library, University of Montana.

25. Grinnell cited by Diettert at page 123.

26. *New York Times* (April 12, 1938), at page 23; *New York Herald Tribune* (April 12, 1938), at page 14.

27. The telegram of April 12, 1938, from E.T. Scoyen to Mrs. George Bird Grinnell is on file at the K. Ross Toole Archives, Mansfield Library, University of Montana.

Epilogue: The Last Stand—"Something Unprecedented"

1. Gerrity, the president of the American Prairie Foundation, is quoted from an article by Hal Herring, "Prairie Dreaming," *Orion* (September/October 2006), at page 57.

2. Scientists disapprove of the term *genetically pure* because they cannot examine every gene in a bison but rather check only some genes when testing for cattle lineage. Scientists instead refer to bison that have a very high probability of having no cattle genes.

3. According to the National Park Service, the fourteen original Wind Cave buffalo were delivered in 1913 by the National Bison Society from stock at the New York Zoological Gardens. (The original New York Zoo stock came from Yellowstone.) In 1916, six more buffalo were delivered to Wind Cave directly from Yellowstone. See the National Park Service timeline at http://www.nps.gov/archive/wica/WICA_Time_Line.htm.

4. Information about the effort by American Prairie Foundation and World Wildlife Fund to restore the buffalo to the high plains of Montana can be found at www.americanprairie.org and www.worldwildlife.org.

5. For an extensive discussion of the use of the buffalo as a symbol, see David A. Dary, *The Buffalo Book* (1989 edition), at chapter XIX, "The American Buffalo as a Symbol."

6. It is estimated that nearly a quarter million buffalo are owned by private ranchers. Significantly, though, virtually all of these animals carry cattle blood, the legacy of decades of efforts to produce a successful cattle–buffalo hybrid. An important issue in relation to domesticated herds was raised by Dale Lott. He called selective breeding the "essence of domestication." The impact of such breeding, he feared, is to "take the wild out" of buffalo. Lott went so far as to claim that the threat of selective breeding is as grave as the "hide hunters' Sharps rifle." Though Lott allowed for the possibility that domestic and wild herds could coexist under the appropriate circumstances, he maintained that "[t]he most vivid threat today is eradication by modification." See Lott, *American Bison* (2002), at pages 192–201. The numbers of Canadian buffalo are from www.thecanadianencyclopedia.com. For a recent examination of the issues raised by domesticated bison, see Curtis H. Freese et al., "Second Chance for the Plains Buffalo," publication forthcoming in *Biological Conservation*.

7. Herring at page 52.

8. Information on the American Prairie Foundation and its prairie project was collected by the author during a visit to their facility south of Malta, Montana, in June 2006. For general information see also www.americanprairie.org.

9. Gerrity is quoted by Hal Harring, at page 57.

Index